Nelson Advanced Science

Mechanics and Radioactivity

r

Mark Ell

THI

First published in 2000 by:
Nelson Thornes Ltd
Delta Place
27 Bath Road
CHELTENHAM
GL53 7TH
United Kingdom

This edition published in 2003

05 06 07 / 10 9 8 7 6 5 4

A catalogue record for this book is available from the British Library

ISBN 0 7487 7660 5

Illustrations by Hardlines and Wearset
Page make-up by Hardlines and Wearset

Printed in Croatia by Zrinski

Contents

Introduction

This series has been written by Principal Examiners and others involved directly with the development of the Edexcel Advanced Subsidiary (AS) and Advanced (A) GCE Physics specifications.

Mechanics and Radioactivity is one of four books in the Nelson Advanced Science (NAS) series developed by updating and reorganising the material from the Nelson Advanced Modular Science (AMS) books to align with the requirements of the Edexcel specifications from September 2000. The books will also be useful for other AS and Advanced courses.

Mechanics and Radioactivity provides coverage of Unit 1 of the Edexcel specification.

Other resources in this series

NAS Teachers' Guide for AS and A Physics provides a proposed teaching scheme together with practical support and answers to all the practice and assessment questions provided in *Mechanics and Radioactivity*; *Electricity and Thermal Physics*; *Waves and Our Universe*; and *Fields, Forces and Synthesis*.

NAS Physics Experiment Sheets, 2nd edn, by Adrian Watt provides a bank of practical experiments that align with the *NAS Physics* series. They give step-by-step instructions for each practical provided and include notes for teachers and technicians.

NAS Make the Grade in AS and A Physics is a revision guide for students. It has been written to be used in conjunction with the other books in this series. It helps students to develop strategies for learning and revision, to check their knowledge and understanding and to practise the skills required for tackling assessment questions.

Features used in this book

The Nelson Advanced Science series contains particular features to help you understand and learn the information provided in the books, and to help you to apply the information to your studies.

Acknowledgements

John Warren and David Hartley painstakingly read through and commented in detail on the manuscripts of the first edition. The authors and publishers gratefully acknowledge their major contributions to the success of the whole series.

Mark Burton and Francis Kirkman gave kind assistance during the preparation of the new Edexcel specification and advised on the provision of Assessment questions.

Photographs

Alan Thomas: 3.3, 3.5;
Allsport: page 11 Simon Bruty, 26.3 Bernard Assett, Vandystadt, 5.2, 20.4;
Image State: 30.4, 31.2;
LAT Photographic: 30.5;
Nelson Thornes: 1.1, 1.6, 1.7, 1.8, 1.9, 3.4, 5.3, 6.1, 6.2, 6.3, 7.1, 8.1, 12.2;
Peter Gould: 20.5, 26.1;
Robert Aberman: 2.4, 34.4;
Rover MG: 6.5;
Science Photolibrary: cover David Parker, ESA, CNES, ArianeSpace, page 1, page 67, 32.1 Phillippe Plially, Eurelios, 3.1 Michael Gilbert, 3.2 Space Telescope Institute, 10.1 Mehau Kulyk, 12.1, 13.4 NASA, 15.6 US Library of Congress, 25.3 Dr Harold Egerton, 26.5 European Space Agency, 27.2 Reverend Ronald Royer, 33.1 Professor Peter Fowler, 33.4 Erich Schrempp

About the authors

Mark Ellse is Principal of Chase Academy in Cannock, Staffordshire, and a former Principal Examiner for Edexcel.

Chris Honeywill is an Assistant Principal Examiner for Edexcel and former Head of Physics at Farnborough Sixth Form College.

Physics: the science of measurement

From smaller than the nucleus of an atom, to the giant scale of galaxies, from events that take place in less than a billionth of a second, to the age of the universe: measurement is central to the study of physics. Measurement allows us to study the properties of different objects and to compare properties between objects. In order to discuss and to compare our measurements universally, we must apply a standard system of units to our measurements.

Measuring the distance to the Moon by sending a laser beam through a telescope and back

1 Making measurements precisely

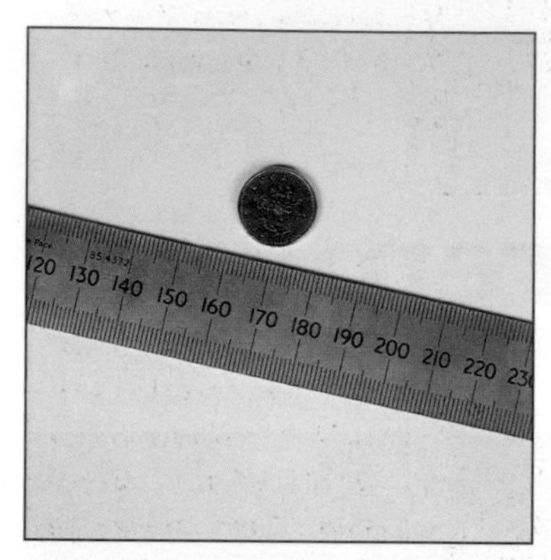

Figure 1.1 The rule can measure to a resolution of 1 mm

Measurements are an important part of your physics course. Here are some instruments for measuring length, and some things to know about all measuring instruments.

Measuring instruments

The **resolution** of a measuring instrument is the smallest difference it can measure, usually the smallest scale division. The scale divisions on the metre rule shown in Figure 1.1 are 1 mm apart. The diameter of the 5p coin it is measuring is 18 mm to a resolution of 1 mm.

Vernier scales are used to divide a wide range of scales into smaller divisions and so provide better resolution. They are commonly used to divide a scale with millimetre divisions into tenths, and so provide a resolution of a tenth of a millimetre (0.1 mm). Figures 1.2 to 1.5 show how to use **vernier callipers**.

The **micrometer screw gauge** uses a screw thread to measure distances to a resolution of a hundredth of a millimetre (0.01 mm).

Figures 1.6 to 1.9 show you how to use the micrometer screw gauge to measure the diameter of a 5p coin.

Uncertainty and precision

No reading of a physical quantity is exact. All readings have some **uncertainty** the range around the measured reading within which the true reading lies. If you read a measuring instrument to one scale division, the uncertainty is at least half a scale division. For instance, an accurate metre rule has an uncertainty of ± 0.5 mm, because a reading could be up to 0.5 mm above or below a scale division and still be quoted as being on that division.

Often the uncertainty depends on the measurement being taken. For instance, if you use a stopwatch, you know that there will be an uncertainty of at least 0.01 s if the stopwatch measures to a hundredth of a second. However, if you are timing a ball falling from the table to the floor, your own reaction time affects the measurement. You might get a range of times, say from 0.34 s to 0.38 s with an average value of 0.36 s, but you are not sure where within this range the true answer is. In this case you would say that the time is 0.36 s ± 0.02 s. The uncertainty is ± 0.02 s. A **precise** measurement has a low uncertainty.

Percentage uncertainty

An uncertainty of 0.5 mm is not important in measuring the height of a building, but would be significant in measuring the diameter of a pencil. You can measure the importance of an uncertainty by dividing the uncertainty by the reading and expressing the result as a percentage:

$$\text{percentage uncertainty} = \frac{\text{uncertainty}}{\text{average value}} \times 100\%$$

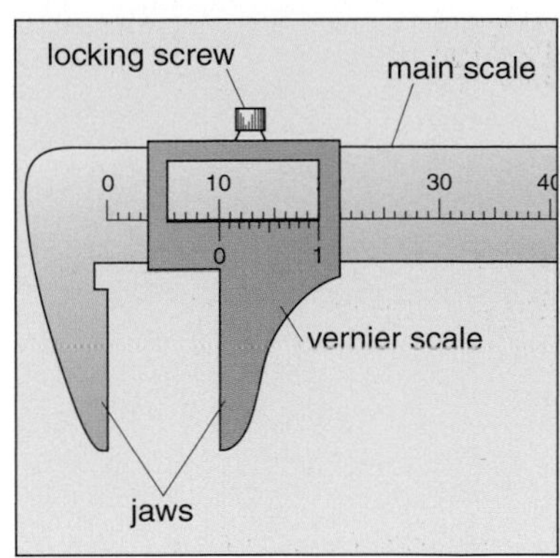

Figure 1.2 *The zero on the vernier scale is at the same end of the scale as the zero on the main scale. The vernier scale has ten divisions in a length of 9 mm, which means that only one of its marks coincides exactly with a mark on the main scale*

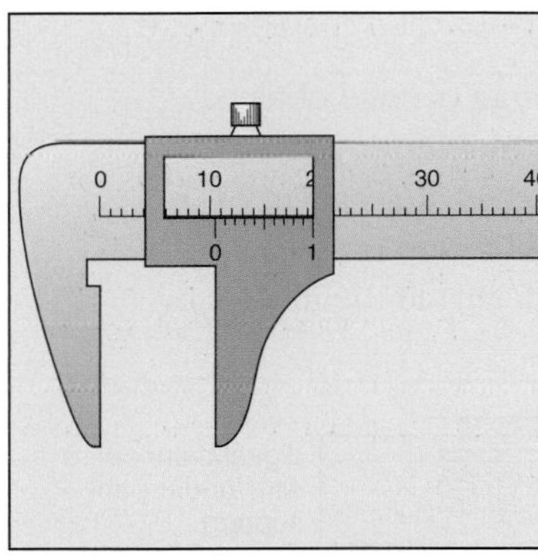

Figure 1.3 *Close the callipers and check that the two zeros coincide. Then open the callipers 10 mm and check that the only mark on the vernier that coincides exactly with the main scale is the zero on the vernier scale. Then ease the callipers half a millimetre further open to 10.5 mm. The zero on the vernier scale is half-way between the 10 mm and 11 mm marks, and the fifth mark on the vernier scale coincides with the mark above it on the main scale*

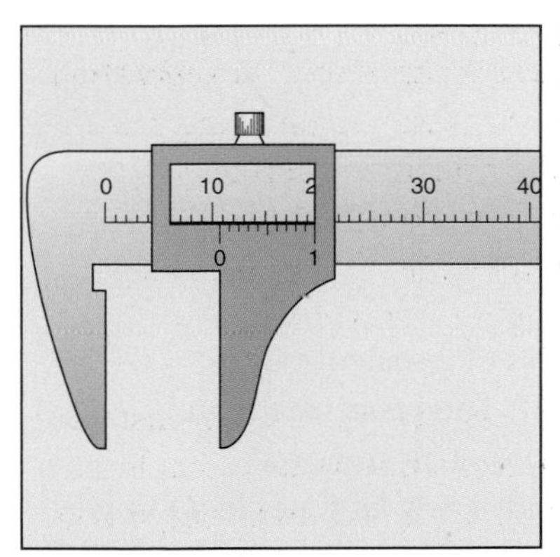

Figure 1.4 *Here the callipers are open 10.7 mm; the seventh division on the vernier scale coincides exactly with the main scale*

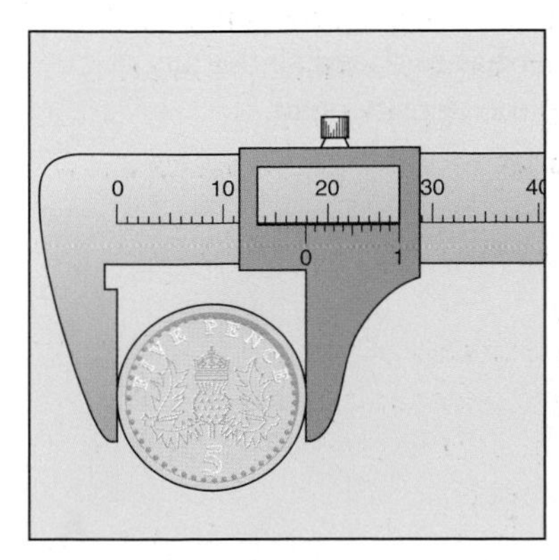

Figure 1.5 *Place a 5p coin in the callipers. First take the reading on the main scale. It is between 17 mm and 18 mm. Then read the vernier scale to find the number of tenth-millimetre divisions – in this case nine tenths, 0.9 mm. So the diameter of the coin is 17.9 mm to a resolution of 0.1 mm*

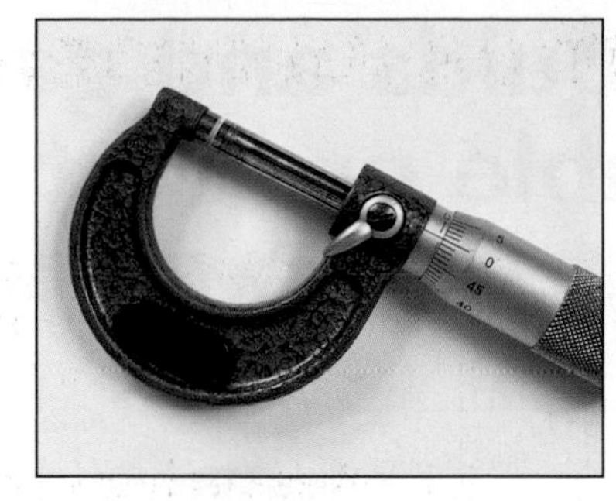

Figure 1.6 *Firstly close the micrometer and check that the scale on the rotating barrel reads zero. Then unscrew the barrel two complete turns. You will find that the gap is 1 mm*

Figure 1.7 *Reduce the gap to 0.5 mm by screwing the barrel back one turn. The first 0.5 mm mark is just visible and the scale on the barrel reads zero. The jaws move 0.5 mm for each turn of the screw and the rotating scale divides this 0.5 mm into 50 divisions, each of 0.01 mm*

Figure 1.8 *Rotate the screw half a turn outwards. The gap is 0.5 mm (shown on the fixed scale) plus 0.25 mm (shown on the barrel), which is a total of 0.75 mm*

Figure 1.9 *Place a 5p coin in the micrometer. Add the fixed scale reading (17.5 mm) to the barrel reading (0.39 mm) to find the diameter of the coin (17.89 mm) to a resolution of 0.01 mm*

2

Solids, liquids and gases: some measurable properties

Measuring density

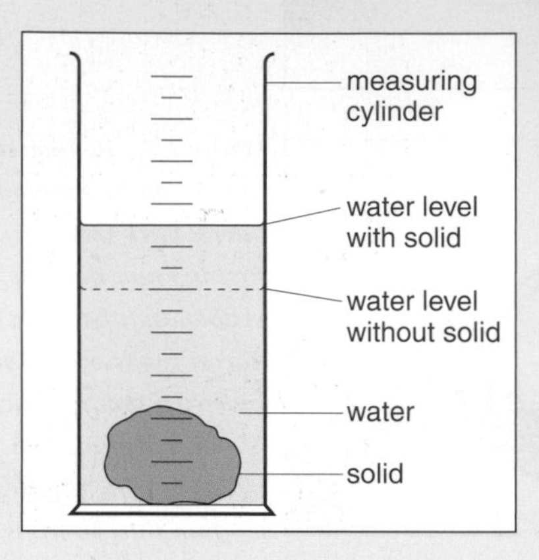

Figure 2.1 Immerse an irregular solid in a measuring cylinder of liquid. The increase in scale indicated by the changing liquid level is equal to the volume of the solid

- Measure the masses in kilograms of a range of solid objects. Then measure their volumes. Use a micrometer or vernier callipers to measure their dimensions if they are regular, or use the method shown in Figure 2.1 if the solids are irregular. Calculate their volumes in metres cubed. Divide the masses by the volumes to find the densities.

- Measure the mass of a measuring cylinder; then add a known volume of liquid to it and measure the total mass. Find the mass of liquid added. Divide the mass of the liquid by its volume to find the density of the liquid.

- Find the density of a cube of ice. Then watch it melt into a measuring cylinder and measure its density again. Carry out a similar experiment with a cube of wax. How much, in general, do substances change in volume when they change from solids to liquids?

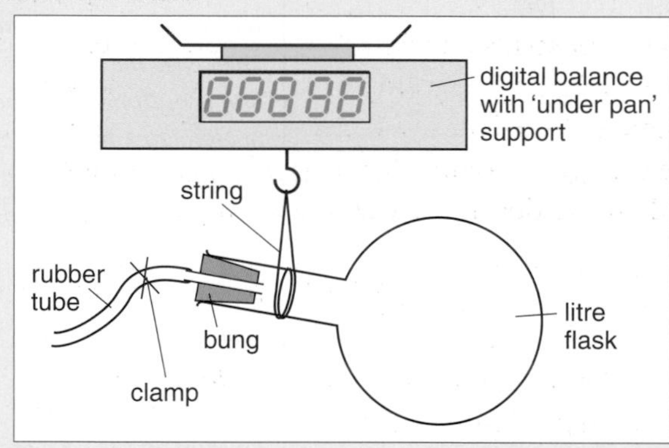

Figure 2.2 Mass of flask includes mass of air inside

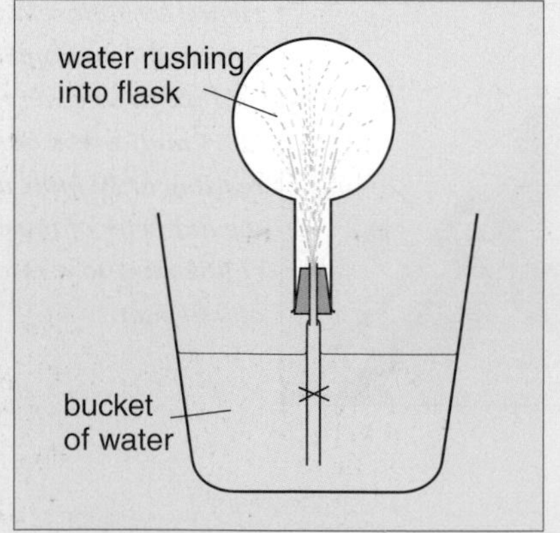

Figure 2.3 Atmospheric pressure pushes water into the empty flask

- Use a digital balance to measure the mass of a sealed round-bottomed flask filled with air (Figure 2.2). Then evacuate it and measure its mass again. Calculate the mass of air removed. Immerse the tube in a bucket of water and slowly let water in. Observe the water rushing into the flask (Figure 2.3).
  Use a measuring cylinder to measure the volume of water that has entered the flask. Assuming that the water has replaced all the air removed by the vacuum pump, calculate the density of air.

Density

The **density** of an object is its mass divided by its volume:

$$\text{density} = \frac{\text{mass}}{\text{volume}} \quad \text{or} \quad \rho = \frac{m}{V}$$

(The symbol ρ is a Greek letter *rho*, pronounced as in 'row the boat'.)

Common, and convenient, units for density are grams per centimetre cubed and kilograms per litre. But the SI unit (see Chapter 3) of density is kilograms per metre cubed (kg m^{-3}), which is sometimes hard to think about unless you remind yourself of how big a metre cubed is (Figure 2.4).

Figure 2.4 Here is a metre cubed. Estimate the volume of the child

The density of water is 1000 kg m^{-3}, and since 1000 kg is a tonne (t), you could express this as 1 t m^{-3}.

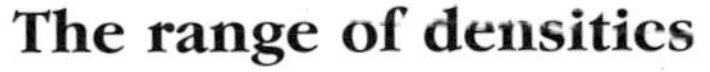

The range of densities

Table 2.1 shows the densities of a range of substances. As you can see, the densities of solids and liquids overlap, and there is no simple rule for which is denser. Solids change little in volume when they are melted, so for a given substance, the densities of the solid and liquid are similar. But there is a very large change in volume from liquid to gas, so the density of any gas is much less than the density of any solid or liquid.

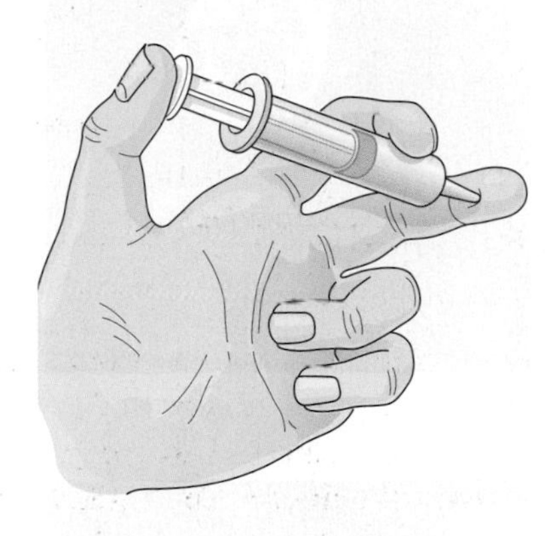

Figure 2.5 Compressing air in a syringe

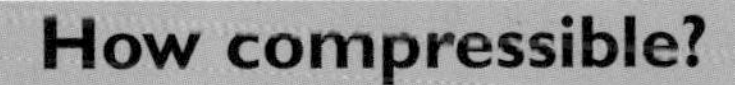

How compressible?

- Try to compress a plastic syringe of air (Figure 2.5).
- Repeat with water.
- Then put a piece of dowel into a syringe and try to compress that.
- Compare the compressibilities with the densities.

Table 2.1 *Densities of some substances*

Substance	Density/kg m^{-3} at room temperature and pressure
platinum	21 400
mercury	13 600
iron	8000
aluminium	2700
glass	2500
water	1000
ice	917 (at –4 °C)
ethanol	790
wood (oak)	c. 700
wood (balsa)	c. 200
carbon dioxide	1.9
air	1.3
methane	0.7
hydrogen	0.09

Compressibility

A fairly small force, which has no noticeable effect on a liquid or a solid, will cause a comparatively large reduction in the volume of a gas. This, along with the fact that the densities of solids and liquids are much greater than those of gases, gives some idea of the microscopic differences between solids and liquids, on the one hand, and gases, on the other.

Rigidity and fluidity

Solids are **rigid**; they keep their shape. You can use solid objects to transmit forces, along a series of levers, for instance.

Gases and liquids are **fluid**. They flow to fit the shape of their container.

3 The International System of Units

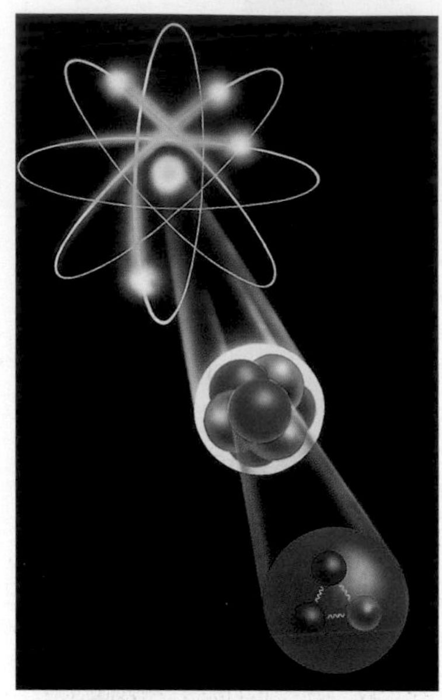

Figure 3.1 Nucleus of an atom

Figure 3.2 A distant galaxy

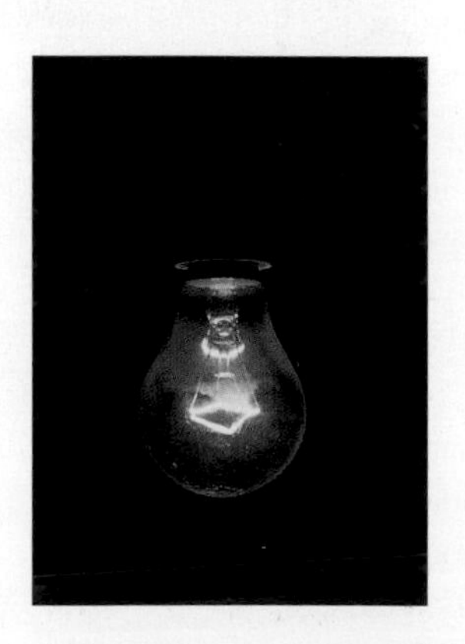

Figure 3.4 The current through this lamp is about a quarter of an amp

Physical quantities

In your study of physics you will measure a range of quantities, from such obvious ones as distance and time to more subtle ones like magnetic flux density and capacitance. All these things that physicists measure are called **physical quantities**.

When studying part of an atom (Figure 3.1), you might be interested in the quantities *mass* and *diameter*. When studying a star (Figure 3.2), you might want to know about the quantities *temperature* and *age*. In an electrical circuit, you will want to know about the quantities *current*, *voltage* and *resistance*.

Base quantities

In any system of units, you have to start from somewhere. These starting quantities are **base quantities**. By international agreement, there are seven base quantities. You will use only six of them in this physics course. These are mass, length, time, current, amount of substance and temperature interval. (The seventh base quantity is luminous intensity.) All other quantities are derived from these base quantities.

Figure 3.3 The mass of an adult male is about 70 kg

Units

If you want to compare your own measurements with others, you need to take measurements in the same way and use the same agreed size or magnitude of quantity for measurement. The agreed magnitudes are called **units** and there are standard units for the whole range of physical quantities.

For measuring the quantity length, the agreed magnitude for comparison (the unit) is the metre. For time, the unit is the second. You may be familiar with the ampere (amp) and the ohm as units of current and resistance. There are less familiar units such as the tesla and the farad as units of magnetic field strength and capacitance.

Base units

The units for the base quantities are defined without reference to any other units. These are **base units**. The International System of Units, abbreviated to SI (from the French *Système International*), agrees base units for the base quantities. In Britain, the National Physical Laboratory works with other laboratories overseas to make sure that base units used in one place are the same as those used elsewhere, so that, for instance, the kilogram used on one balance is the same as the kilogram used on any other.

Table 3.1 lists the base units along with their symbols. Most base units are defined in a way that can be reproduced anywhere in the world, but the kilogram is, rather inconveniently, defined by a standard kilogram (called the *prototype kilogram*) kept in Paris. This means that all other kilogram masses must be compared, directly or indirectly, with the prototype.

Table 3.1 *Base quantities and units*

Base quantity	Base unit	Symbol	Definition
length	metre	m	the distance an electromagnetic wave travels in a time of 1/299 792 458 second
mass	kilogram	kg	the mass equal to that of the standard kilogram
time	second	s	9 192 631 770 oscillations of a caesium atomic clock
current	ampere	A	that constant current which, when it is flowing in two infinitely long parallel thin wires that are 1 m apart in vacuum, produces between them a force per unit length of 2×10^{-7} N m^{-1}
temperature interval	kelvin	K	1/273.16 of the thermodynamic temperature of the triple point of water
amount of substance	mole	mol	the amount of substance that contains as many elementary units as there are atoms in 12 g of carbon-12

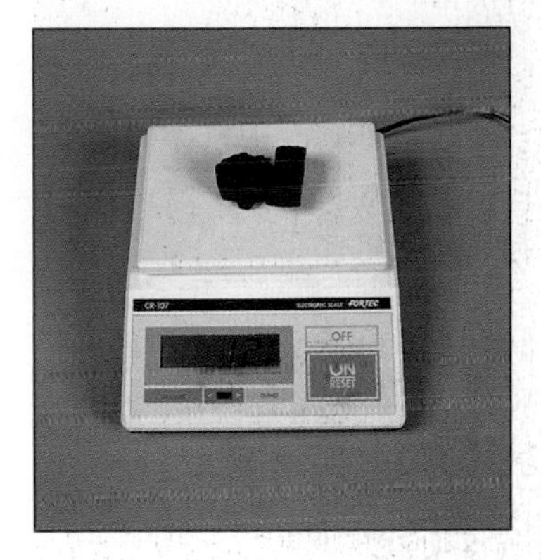

Figure 3.5 Twelve grams of carbon contain one mole of atoms

Measurements and units

You express physical measurements as multiples of a physical unit. For instance, a race track might be 100 metres long – 100 times as long as the metre, the standard unit for length. A girl's mass might be 55 kg – 55 times as large as the mass standard, the kilogram. All measurements are expressed like this, as a number multiplied by a unit. Larger and smaller multiples can be written with prefixes (Tables 3.2 and 3.3). The measurement is the *product* of a number and a unit. You need to include both the number and the unit when stating the result of a measurement. If you leave either out, then the statement is meaningless. Figures 3.3 to 3.5 show some examples of physical measurements.

Table 3.2 *You can use prefixes with units to express larger or smaller multiples*

Prefix	Symbol	Multiplier
giga	G	10^9
mega	M	10^6
kilo	k	10^3
milli	m	10^{-3}
micro	μ	10^{-6}
nano	n	10^{-9}
pico	p	10^{-12}

Table 3.3 *Some typical magnitudes of base quantities*

	Length/m	Mass/g	Time/s	Current/A	Temperature/K	Amount of substance/mol
pico (10^{-12})	wavelength of a gamma ray	10^{12} protons	period of molecular spin	current when 6×10^6 electrons pass a point in one second		10^{10} units
nano (10^{-9})	length of a molecule of olive oil	speck of dust	very fast computer switch time	ionisation current caused by a match flame		amount of carbon in this full stop.
micro (10^{-6})	typical size of a bacterium	small grain of sand	period of a radio wave	current drawn by a digital watch	about as cold as you can get	grain of salt
milli (10^{-3})	diameter of a pinhead	raindrop	period of a sound wave	current through 1.5 k resistor connected to a dry cell		carbon dioxide exhaled during one breath
reference unit	height of a laboratory bench	smallest mammal (1 g)	tick of a grandfather clock	current drawn by a car headlamp	cosmic background temperature	2 g of hydrogen gas
kilo (10^3)	Big Ben to Nelson's column	1 litre of milk (1 kg)	mean lifetime of a free neutron	current drawn by a 'Chunnel' train	melting point of silver	bucketful of water
mega (10^6)	Land's End to John O'Groats	1 m^3 of water(1000 kg = 1 tonne)	11 days	current drawn by all the houses in a large town	outermost atmosphere of the sun	your annual consumption of water
giga (10^9)	diameter of the sun	large tree from a rainforest (1000 t)	duration of the Thirty Years' War	current drawn by all the towns in Britain	core of the hottest stars	oxygen inhaled during a lifetime

4 Derived quantities and units

Derived quantities

Many quantities, like speed, force, density, charge etc, are not base quantities. They are *derived* from base quantities. So they are called **derived quantities**. Derived quantities are defined by word equations. For example, the derived quantity speed is defined in terms of the two base quantities distance and time by the equation

$$\text{speed} = \frac{\text{distance}}{\text{time}}$$

All derived quantities are built up from base quantities by a series of word equations. You can use an equation with symbols instead of a word equation if you state the meaning of the symbols. For example, $v = x/t$, where v is speed and x is distance covered in time t.

Derived units

Derived units are defined with reference to the base units. The unit of force, the newton, is a derived unit. It is defined from the base units kilogram, metre and second.

The word equation that defines a derived physical quantity also defines its unit. For example, the units of distance and time are metres and seconds, so the unit of speed is the metre per second.

Homogeneity

An equation which states that two things are equal makes sense only if those two things are of the same type. It makes sense to say

$$5\ \text{kg} = 3\ \text{kg} + 2\ \text{kg}$$

but no sense at all to say

$$5\ \text{kg} = 3\ \text{kg} + 2\ \text{m} \qquad \text{or} \qquad 5\ \text{kg} = 5\ \text{m}\,\text{s}^{-1}$$

Even though the numbers are equal, the units show that the quantities are not of the same type.

Quantities that you say are equal, or that you add together, must be of the same type. The word **homogeneous**, from Greek, means *same type*. A physical equation can be correct only if it equates or adds together homogeneous quantities.

Correct equations must be homogeneous, but homogeneous equations are not automatically correct. Even if the units are equivalent, the numbers may not be correct. The equations

$$5\ \text{kg} = 3\ \text{kg} + 1\ \text{kg} \qquad \text{or} \qquad \text{speed} = 6 \times \frac{\text{distance}}{\text{time}}$$

are both homogeneous, but clearly incorrect.

Mechanics

To withstand the force of an earthquake, engineers have learned how to build structures that move, rather than crumble. Road safety experts study the effects of collisions to enhance safety aspects of cars. Both of these examples involve *mechanics* – the study of how forces affect objects. Mechanics helps us predict the outcome of collisions and calculate the amount of energy required to launch the Space Shuttle. Indeed, everything we do, every day, is influenced by mechanics.

An example of 120 kg moving at 10 m s^{-1}

5 Displacement and velocity

Distance and displacement

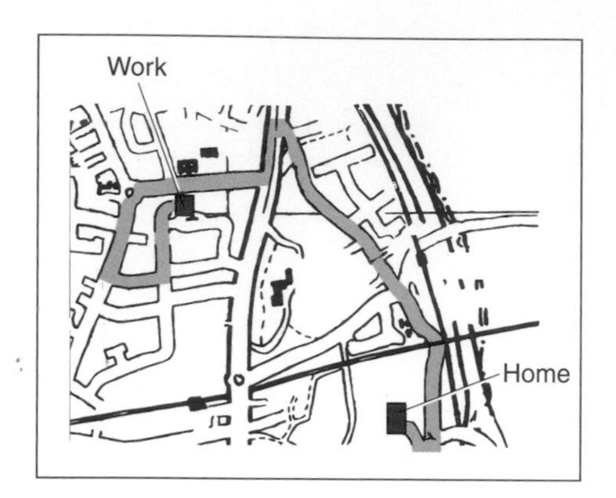

Figure 5.1 Though the distance travelled from home to work is 4 km, the displacement from home is never as much as 2 km

The study of motion is called kinematics. The easiest thing to measure about motion is how far something moves: the distance it travels.

If you walk 4 km from home to work, and 4 km back from work to home, then you have travelled a total distance of 8 km. On the other hand, by the end of the day you are back where you started, so you end up no distance at all from your starting point. If your route to work has corners and bends, you will never be as much as 4 km from your home. Distance to work alone does not give you information about how *far* it is from home to work. Nor does distance indicate the *direction* from home to work.

There is another quantity that states both the magnitude and direction of a change in position. It is called **displacement** – the distance moved in a particular direction. If your route to work is like that in Figure 5.1, your displacement during the day is about 2 km north-west. Half-way between home and work, the displacement is still about 2 km, but in a different direction; when you get back home at night, your displacement is zero.

Vectors and scalars

Distance and displacement are examples of two different types of quantities: **scalars** and **vectors**.

- Scalar quantities have size (magnitude) only and no direction. All the base quantities are scalar quantities and, like distance, have no direction.

- Vector quantities have both size and direction. Displacement has both size and direction, and is a vector quantity. Other examples of vector quantities are force, velocity and acceleration.

Speed

Figure 5.2 A sprinter increases speed quickly over the first few metres

Speed is distance moved per second. You find it by measuring the time to travel a known distance and then dividing the distance by the time. That is,

$$\text{speed} = \frac{\text{distance}}{\text{time}}$$

In many situations, the speed of an object is changing. For example, a sprinter (Figure 5.2) starts slowly and quickly gathers speed. The **average speed**, as its name implies, averages all the different speeds:

$$\text{average speed} = \frac{\text{total distance travelled}}{\text{total time taken}}$$

Both speed and average speed are scalar quantities; direction is not relevant.

Velocity

Just as displacement is the vector that corresponds to distance, **velocity** is the vector that corresponds to speed:

$$\text{velocity} = \frac{\text{displacement}}{\text{time}}$$

As with displacement, strictly you should give the direction of any velocity. A car that is travelling northwards along a road at a *speed* of 30 m s^{-1} has a *velocity* of 30 m s^{-1} *northwards*.

A car moving along a road at a steady speed changes direction as the road bends. Though its speed is constant, its direction of motion is not; so the velocity is changing as it travels around the bend.

Many kinematics problems concern linear motion, in which objects move only in straight lines. In these, the direction is often obvious and you do not need to specify it all the time. But objects moving in a straight line go backwards as well as forwards, so you do need to specify one direction as positive, and use positive and negative signs to show direction.

Measuring velocity

- Set up a sloping runway and measure its length. Release a trolley from the top and time how long it takes to run to the bottom.
- Calculate the average velocity.
- Repeat your measurements and calculations a few times to check that they are reliable.
- Put a card of measured length on top of the trolley and use a timer to measure how long the card takes to pass through a light beam. (Figure 5.3)
- Measure the velocity of the trolley like this near the top of the runway and near the bottom of the runway. (You may have an intelligent timer that does the calculations for you.)
- Compare these velocities with the average velocity down the runway.

Figure 5.3 Trolley with card interrupter

Average velocity and instantaneous velocity

The average velocity is the total displacement divided by the total time. If the velocity is changing, the velocity at any instant, called the **instantaneous velocity**, is different from the average velocity. To find the instantaneous velocity, you need to measure the change in displacement over a very short time interval, and divide that displacement by the short time.

6 Acceleration

Acceleration

When the velocity of any object changes, the object is accelerating. A trolley running down a steep runway is accelerating. The **acceleration** is the quantity that indicates how fast the velocity is changing. It is the change in velocity per second:

$$\text{acceleration} = \frac{\text{change in velocity}}{\text{time taken for change}}$$

If the trolley starts at rest, and 2 s later has a speed of 1.8 m s^{-1}, then

$$\text{acceleration} = \frac{\text{change in velocity}}{\text{time taken for change}} = \frac{(1.8 \text{ m s}^{-1} - 0 \text{ m s}^{-1})}{2 \text{ s}} = 0.9 \text{ m s}^{-2}$$

As the calculation shows, the units of acceleration are m s^{-2}.

The size of the acceleration indicates how much the velocity changes every second. The trolley starts at rest. One second later its velocity is 0.9 m s^{-1}. Two seconds after starting, its velocity is 1.8 m s^{-1}. What will the velocity be 3 s after starting if the acceleration continues at the same rate?

Acceleration is a vector quantity; direction is important. If you measure displacement down the slope as positive, the trolley on the runway has a positive acceleration; the acceleration causes the velocity down the slope to increase.

Measuring acceleration

- Set up a sloping runway. Fix two cards of the same known length onto the trolley so that they interrupt a light beam twice as the trolley rolls down (Figure 6.1). Let the trolley roll; use an intelligent timer to measure the time of the two interruptions and the time between them.
- Calculate the average velocity of the trolley during each interruption.
- Divide the change in velocity by the time taken for the change, to find the acceleration.
- Drop a double interrupter card through a light gate connected to a timer to measure the acceleration of the falling card (Figure 6.2). What value do you get?
- Add masses to the card and repeat your measurements.

Figure 6.1 Trolley with double interrupter

Figure 6.2 A double interrupter card dropping through a light gate

Using ticker timers and video cameras to measure acceleration

- A ticker timer puts 50 ticks on a tape per second. Let a trolley pull a tape through the ticker timer as it runs down the runway (Figure 6.3).
- Calculate the average velocity at 0.1 s intervals (Figure 6.4) and plot a graph of velocity against time to calculate the acceleration.
- A video camera takes 25 pictures per second. Video your trolley running down the runway, with a calibrated scale in the background. Play the video back a frame at a time and measure from the screen the distance travelled per frame.
- Compare these results with those using the ticker timer.

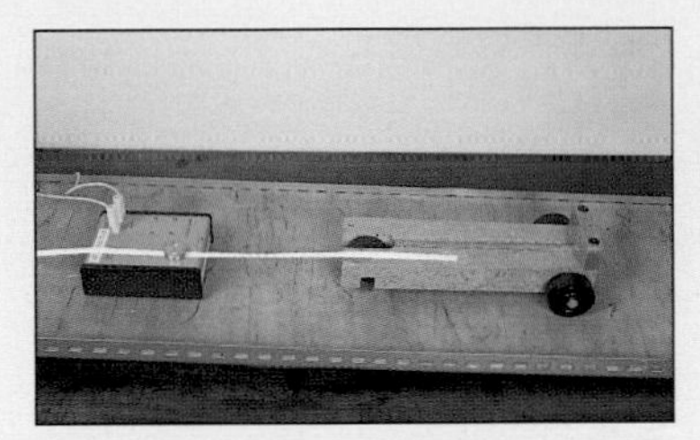

Figure 6.3 Trolley with ticker timer

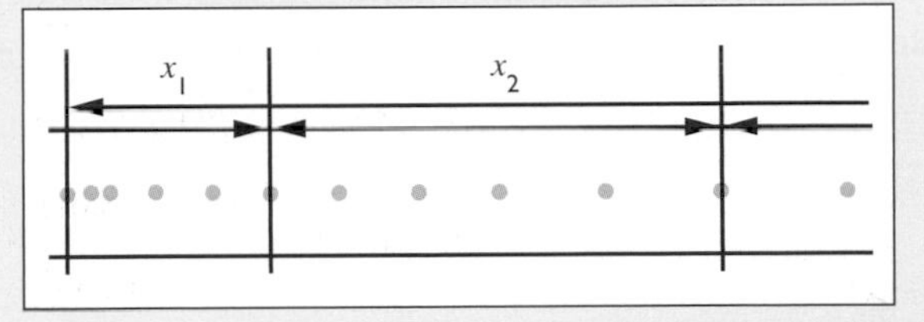

Figure 6.4 Mark the start of the tape (first tick) and tabulate values of the distance every five ticks afterwards. This is the distance covered each tenth of a second

Rate of change

Rate of means 'divided by time'. So rate of change of displacement means change in displacement divided by time, which is velocity. Rate of change of velocity means change in velocity divided by time, which is acceleration.

Figure 6.5 This MG takes about 8 s to accelerate from rest to 30 m/s

Motion graphs

Measuring displacement

- A motion sensor sends out a pulse of ultrasound waves and times how long the pulse takes to return, so it can determine how far away an object is.
- Set up the motion sensor on a runway. Use it to measure how far away from it the trolley is (Figure 7.1). Check its measurements with a ruler.
- Let the trolley run down the slope and use the sensor and computer to plot a graph to show how the displacement changes with time.
- Repeat for a range of different slopes.

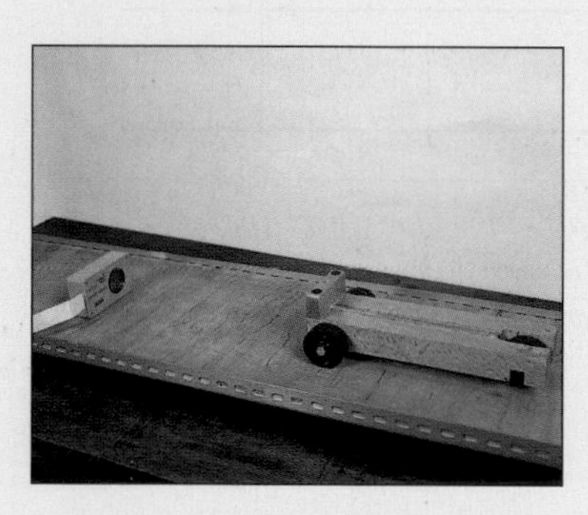

Figure 7.1 A motion sensor can measure distance to the trolley

Displacement–time graphs

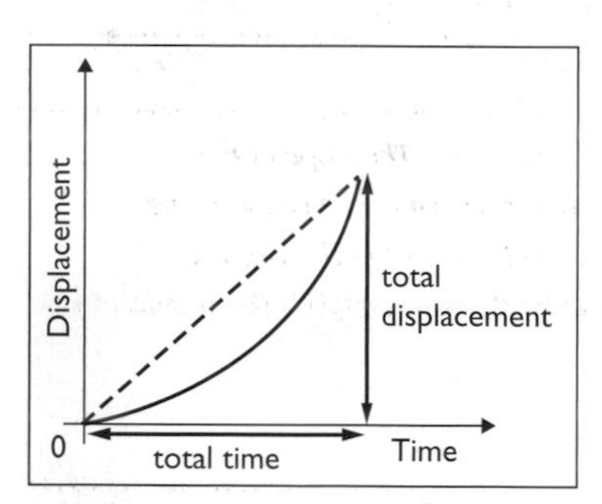

Figure 7.2 The slope of the dotted line is the average velocity

Figure 7.2 shows how the displacement varies with time for the trolley running down the runway. At the beginning, the graph is shallow; at the end it is steeper. The mathematical quantity which measures the steepness of the graph is the gradient. When the graph is a curve, the gradient is changing.
You calculate the gradient of any graph by using the formula

$$\text{gradient} = \frac{\text{change in vertical step}}{\text{corresponding change in horizontal step}}$$

For the graph in Figure 7.2, the total change in the vertical step is equal to the total displacement, and the total change in horizontal step is equal to the time for that displacement. So

$$\text{gradient of the dotted line} = \frac{\text{total displacement}}{\text{time for that displacement}} = \text{average velocity}$$

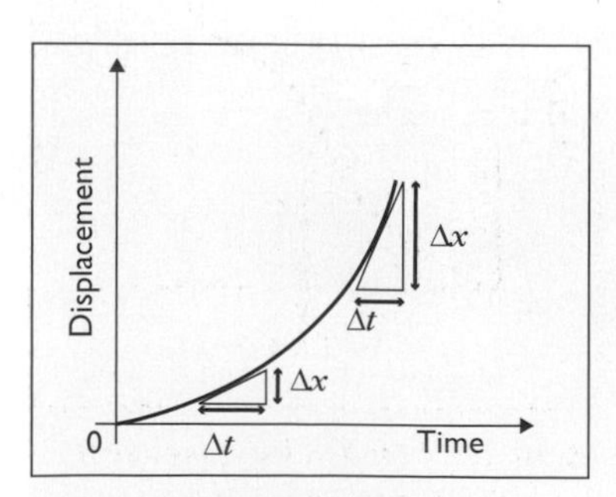

Figure 7.3 The instantaneous velocity at two points

The gradient at any point on the graph is equal to the tiny change in vertical step divided by the corresponding horizontal step. The **gradient of a displacement–time graph** at any point is the instantaneous velocity at that point. As Figure 7.3 shows, the instantaneous velocity of the trolley increases as it runs down the slope.

The symbol Δx is used to represent the small displacements used to calculate the slope, and the corresponding times are indicated by the symbol Δt. The velocity is $\Delta x/\Delta t$. (The Greek letter delta Δ is used to indicate that the changes are small.) The quantity $\Delta x/\Delta t$ at any point is equal to the gradient at that point and is therefore the instantaneous velocity.

By seeing how the slope of a displacement–time graph changes, you can produce a velocity–time graph. Figure 7.4 shows the displacement–time and velocity–time graphs for a trolley running down a runway.

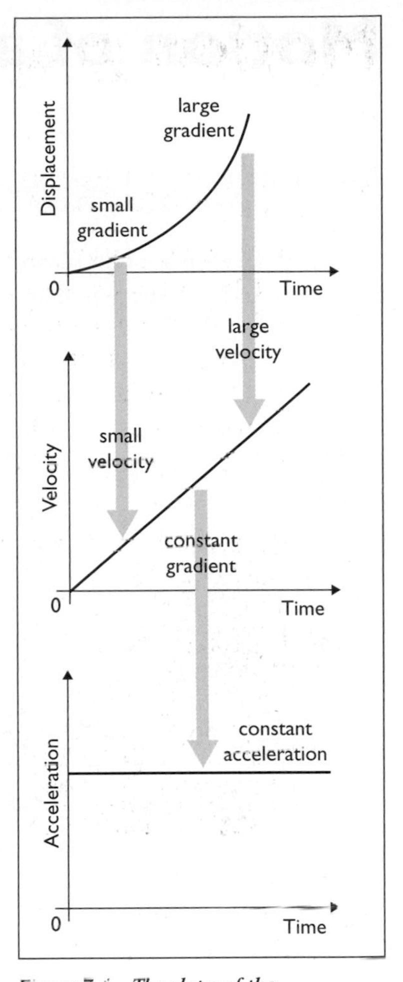

Figure 7.4 The slope of the displacement–time graph is the velocity, and the slope of the velocity–time graph is the acceleration

Velocity–time graphs

A trolley running down a slope starts off at rest and its velocity increases by the same amount every second. The velocity–time graph is a straight line through the origin. The **gradient of a velocity–time graph** is the change in velocity divided by the time for that change. This is the acceleration.

For this graph, the gradient, and therefore the acceleration, is $\Delta v/\Delta t$, where Δv is the change in velocity and Δt is the time taken for that change. (Again the letter Δ is used to indicate that the changes are small.)

For the trolley running down the slope, the velocity–time graph is a straight line. Its gradient is constant, showing that the acceleration is constant, as shown by Figure 7.4.

Area under a velocity–time graph

Figure 7.5 shows a velocity–time graph. The area under the graph for the small time Δt is shaded. Since the strip is narrow, the average velocity is approximately the same as v, the velocity at the midpoint of the narrow strip. The area of the strip is $v\Delta t$, which is also equal to the average velocity multiplied by the time. This is the change of displacement in time Δt.

The same is true for all the rest of the strips that make up the area under the graph. So the final displacement is the total area underneath the graph. For any velocity–time graph, the **area under a velocity–time graph** is the displacement during this time.

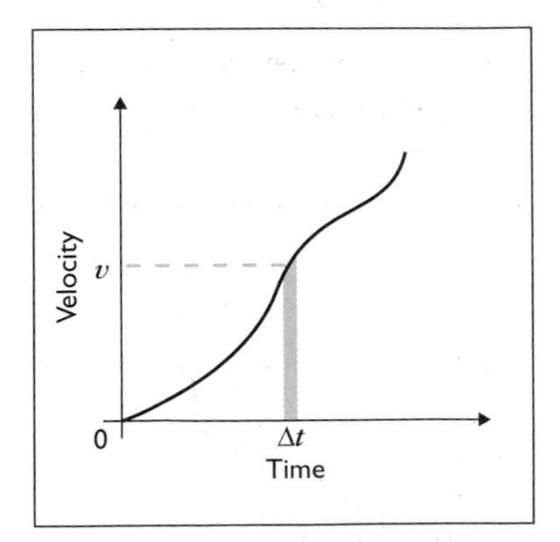

Figure 7.5 The shaded area under the velocity–time graph is the displacement during time Δt

8

Motion of a bouncing ball

Bouncing ball

- Use the motion sensor to produce displacement–time, velocity–time and acceleration–time graphs for a bouncing ball (Figure 8.1).

Figure 8.1 Set up the motion sensor above the bouncing ball

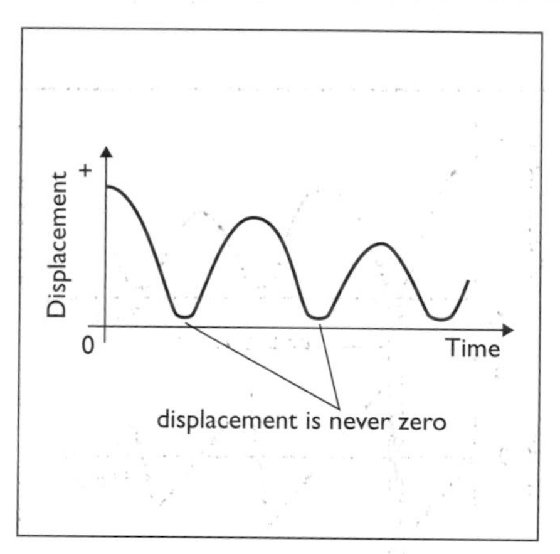

Figure 8.2 Displacement–time graph of the centre of a bouncing ball above the ground

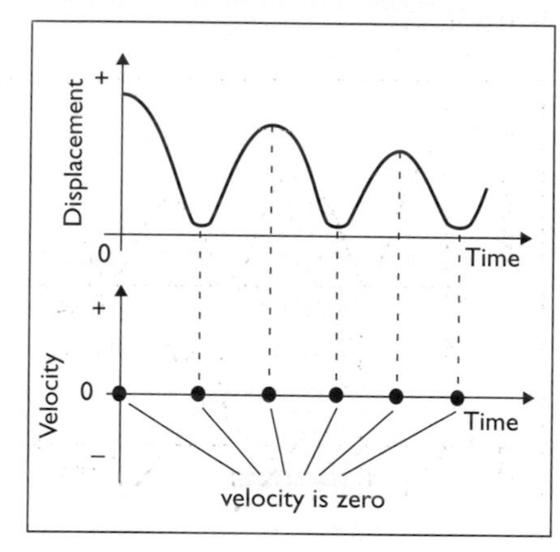

Figure 8.3 The velocity is zero where the gradient is zero

Sketching kinematics graphs

The displacement–time graph for a bouncing ball looks familiar enough (Figure 8.2). But the situation is complicated, and a good one in which to practise sketching velocity–time and acceleration–time graphs from a displacement–time graph. Figures 8.2 to 8.7 guide you through this process.

Notice that the displacement is never zero (Figure 8.2), because the centre of the ball never touches the ground. The velocity is zero where the gradient of the displacement–time graph is zero. First, mark these points (Figure 8.3). At points like A (Figure 8.4), the gradient of the displacement–time graph is large and positive; so is the velocity. At points like B, the slope of the displacement–time graph is large and negative; so is the velocity.

When the ball is in contact with the ground (Figure 8.5), i.e. during the brief bounces, the gradient changes rapidly from negative to positive; so does the velocity.

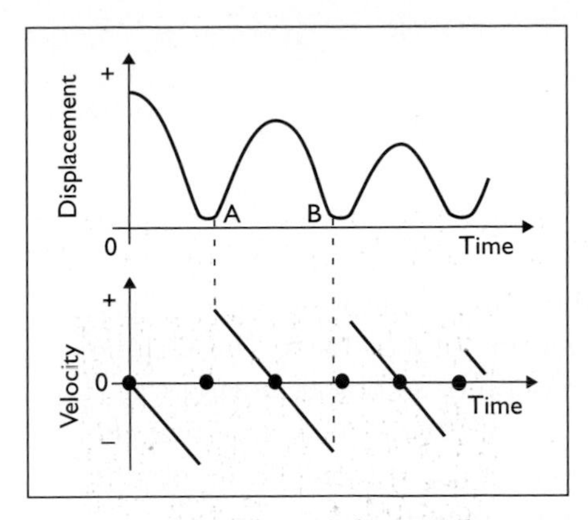

Figure 8.4 The velocity is large and positive at points like A, and large and negative at points like B

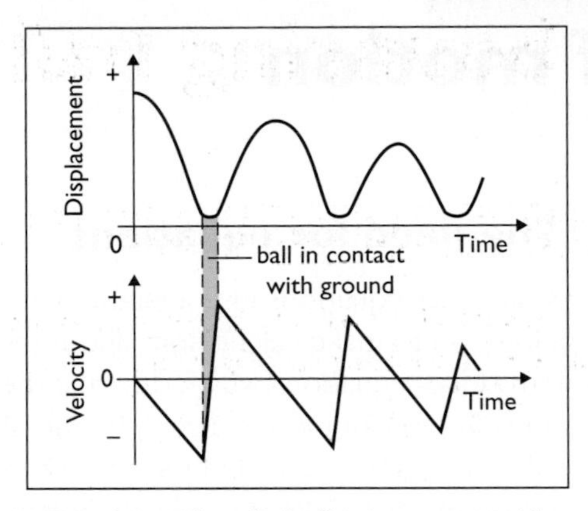

Figure 8.5 When the ball is in contact with the ground, the velocity changes rapidly

While the ball is in the air (Figure 8.6), the gradient of the velocity–time graph is constant and negative; so is the acceleration. The acceleration is constant, because it is equal to the acceleration of gravity. The acceleration is negative because it is downwards and we are using a sign convention that says upwards is positive.

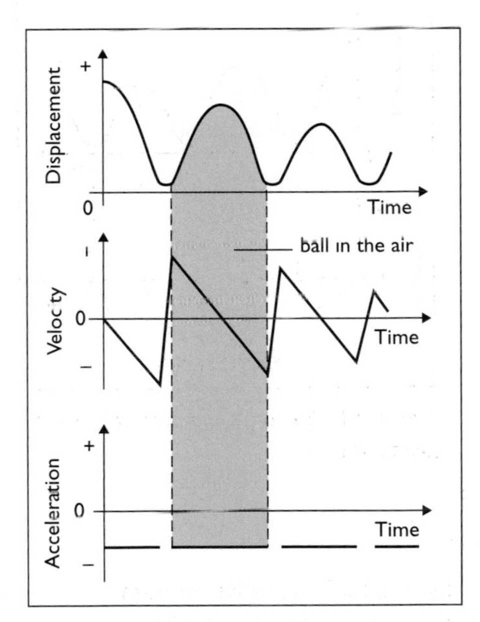

Figure 8.6 While the ball is in the air, the acceleration is constant and negative

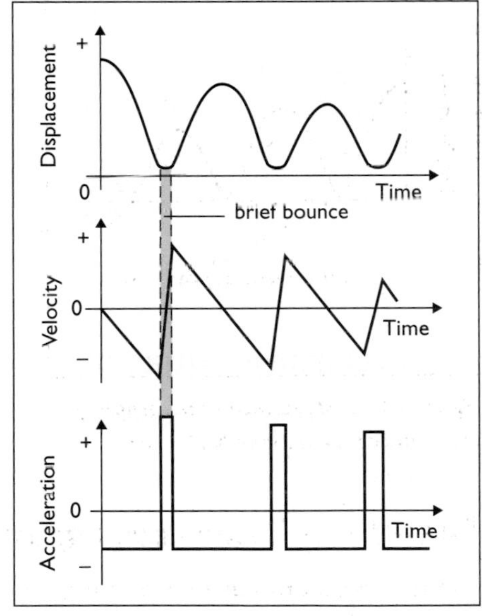

Figure 8.7 During the brief bounces, the acceleration is large and positive

During the brief bounces (Figure 8.7), the velocity changes a large amount from negative to positive as the ball changes direction. This is a large positive change of velocity in a short time. The large and positive acceleration is caused by the force of the ground on the ball. The gradient of the velocity–time graph is large and positive; so is the acceleration. It is greater than the downwards acceleration of gravity.

9 Equations of motion

Table 9.1 *Symbols used in kinematics equations*

v	final velocity	m s^{-1}
u	initial velocity	m s^{-1}
x	displacement	m
a	acceleration	m s^{-2}
t	time	s

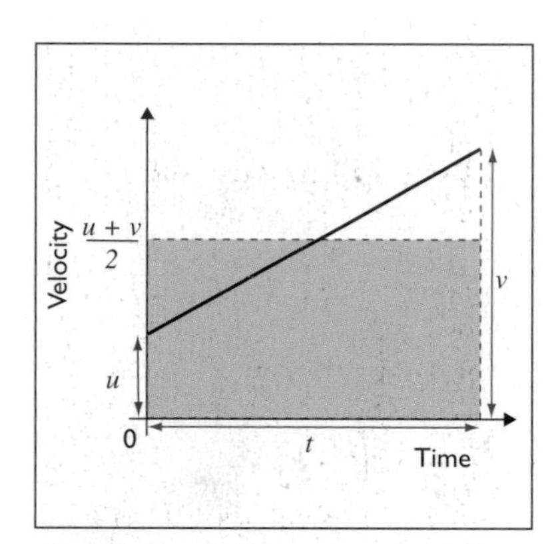

Figure 9.1 Average velocity is $(u + v)/2$

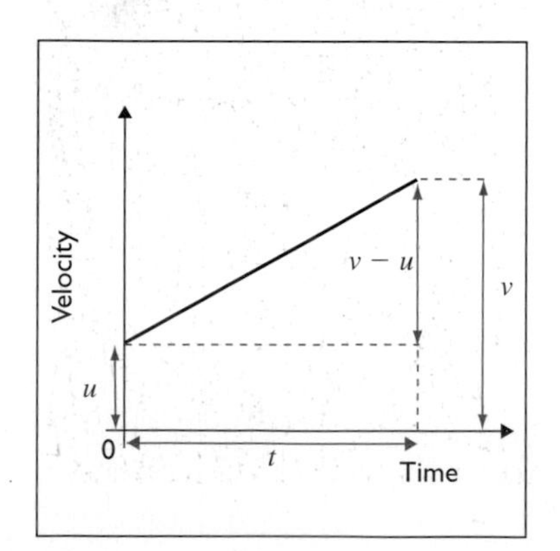

Figure 9.2 Final velocity is $v = u + (v - u) = u + at$

The need for equations

Kinematics equations give a mathematical picture of a moving object. They allow you to make calculations about the motion of the object. This chapter introduces equations which describe the motion of an object moving with constant acceleration in a straight line, the most common situation that you will encounter. The chapter shows where the equations come from, but the important thing is that you know how to apply them.

Table 9.1 shows the symbols and the units used for final and initial velocities, displacement, acceleration and time.

Average velocity equation

You know that:

$$\text{average velocity} = \frac{\text{displacement}}{\text{time taken}}$$

If a body has a constant acceleration, you can find the average velocity by adding together the initial and final velocities and dividing by 2. So:

$$\text{average velocity} = \frac{(u + v)}{2} = \frac{x}{t} \quad \text{Rearranging,} \quad \mathbf{x = \frac{(u + v)}{2}\, t}$$

This is shown in Figure 9.1.

Acceleration equation

The word equation for acceleration is used to derive this equation:

$$\text{acceleration} = \frac{\text{change in velocity}}{\text{time for change}}$$

Since change in velocity = final velocity – initial velocity
you can write

$$a = \frac{(v - u)}{t} \quad \text{giving} \quad at = v - u$$

so $\mathbf{v = u + at}$

This is shown in Figure 9.2.

Displacement equation

Displacement is equal to the area under a velocity–time graph. Figure 9.3 shows a body with initial velocity u. It accelerates at a constant rate to velocity v in a time t. The body's displacement is the area under the graph; it is split up into a rectangle and a triangle.

The rectangle has height u and length t. Its area is therefore equal to ut. The triangle also has a base length t. Its height is $v–u$ and from the acceleration equation, we know that $v–u = at$. We therefore have a triangle of base t and height at. The area of this is $\frac{1}{2}at^2$. So

displacement = total area under graph

= area of rectangle + area of triangle

so $$x = ut + \tfrac{1}{2}at^2$$

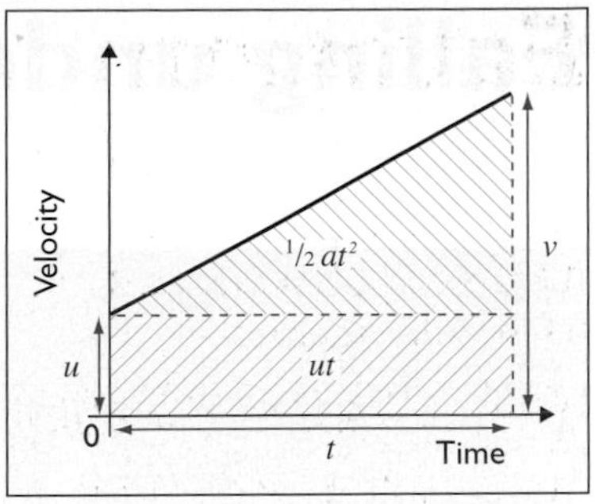

Figure 9.3 Displacement x equals the area under a velocity–time graph, that is, $x = ut + \frac{1}{2}at^2$

The fourth equation

You can combine two of the above equations to make another useful equation which leaves out the quantity time. First rearrange $v = u + at$ to give an equation for t:

$$t = \frac{(v - u)}{a}$$

Then substitute this into the equation

$$x = \frac{(u + v)}{2}\,t$$

This gives

$$x = \frac{(u + v)}{2}\,\frac{(v - u)}{a}$$

$$x = \frac{(u + v)(v - u)}{2a}$$

$$2ax = (u + v)(v - u)$$

$$2ax = v^2 - u^2$$

so $$v^2 = u^2 + 2ax$$

WORKED EXAMPLE

A sprinter starts from rest and reaches a speed of 10 m s^{-1} after covering a distance of 500 cm. Calculate the acceleration, assuming that this is constant.

We have

$v = 10$ m s^{-1}

$u = 0$ since starting from rest

$x = 500$ cm $= 5$ m

$a = ?$ (m s^{-2}) you want to find this

$t =$ (irrelevant)

The equation which contains v, u, x and a is

$$v^2 = u^2 + 2ax$$

Numerically,

$10^2 = 0^2 + 2 \times a \times 5$

$100 = 10a$

$a = 10$ m s^{-2}

Some tips for solving kinematics problems

1. Be careful and methodical in your physics calculations.
2. Remember that kinematics equations apply only when (i) the acceleration is constant, (ii) the motion is in a straight line and (iii) the displacement x is 0 when the time t is 0.
3. Say '*vuxat*' and make a little table of the quantities showing what you know and what you need to find out. Remember that 'starting from rest' means that $u = 0$.
4. Convert the units to a coherent set, usually metres and seconds.
5. Choose your equation to contain the quantity you need and those that you know.
6. Substitute, rearrange and calculate.
7. Remember the unit in your answer.

10 Falling under gravity

Figure 10.1 The hammer and feather experiment

Equal acceleration

Chapter 6 described how to measure acceleration by timing two interruptions of a light gate, by using ticker tape, and by videoing an accelerating object. If you use these methods to measure the acceleration of a range of falling bodies, you find that for most bodies the acceleration is the same. Of course, in air, a very light body like a feather has less acceleration than a denser body like a hammer. But when both fall freely through a vacuum, it is easy to demonstrate that they have the same acceleration (Figure 10.1). This is the acceleration due to gravity, known as g.

For most objects falling through a few metres, from cannon balls to tennis balls, air resistance has little effect and their acceleration in air is constant and equal to g.

Timing a falling ball

- The electromagnet holds the steel ball when the switch is in position A in Figure 10.2.
- Measure the distance x from the bottom of the ball to the top of the trapdoor.
- Switch to position B. The electromagnet releases the ball, the clock circuit is completed and the clock starts.
- When the ball hits the trapdoor, it opens the clock circuit and the clock stops. Record the time t for the ball to fall.
- Repeat and measure the time twice more. Average the three values.
- Repeat for a range of different heights.
- Plot a graph of x against t^2.

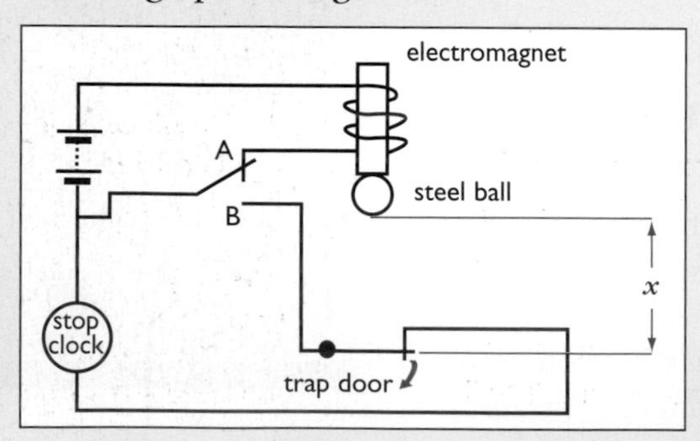

Figure 10.2 Diagram showing circuit for accurate timing of a falling ball

Calculating g from the time taken to fall a known distance

Air resistance hardly affects the acceleration of a falling steel ball, so its acceleration is very close to that due to gravity.

The ball falls a distance x from rest in a time t. The acceleration can then be found from the equation

$$x = ut + \tfrac{1}{2}at^2$$

Since $u = 0$, and the acceleration $a = g$:

$$x = \tfrac{1}{2}gt^2$$

Rearranging:

$$g = \frac{2x}{t^2}$$

If you measure x and t, you can use this equation to calculate a single value for the acceleration.

A better method for measuring g is to make a series of measurements of the time t taken to fall a range of distances x. Compare the equation

$$x = \tfrac{1}{2}gt^2$$

with

$$y = mx + c$$

which is the equation for a straight line with gradient m and intercept c on the y axis. If you plot a graph of x against t^2, its gradient will be $\tfrac{1}{2}g$. The constant c will be zero, showing that the graph will go through the origin. You can measure the gradient of your graph and double it to find g.

Vertical projection

Projection simply means throwing. If you project an object into the air, its acceleration after you have let it go is always downwards, whether it is on the way up, at the top, or on the way down. If air resistance is small, its downwards acceleration always equals g (around 9.81 m s^{-2}).

Now look at the Worked Example. If you hit a cricket ball hard into the air, it may well have an initial speed of 40 m s^{-1}. The calculation indicates that it will reach a height of 82 m. In practice, over such large distances, air resistance is significant. It slows the ball on the way up, so the actual height reached is less than 82 m. In the same way, air resistance affects the flight of a bullet fired into the air, so the bullet does not come down at the same speed at which it left the gun.

WORKED EXAMPLE

You hit a cricket ball vertically upwards at 40 m s^{-1}. How high does it go?

We have

$v = 0$ (at the top of the path)

$u = 40$ m s^{-1} (call upwards positive, so u is positive)

$x = ?$

$a = -9.81$ m s^{-2} (acceleration is downwards, so negative)

t (irrelevant)

so we use

$v^2 = u^2 + 2ax$

Numerically

$0^2 = 40^2 + 2 \times (-9.81) \times x$

$0 = 1600 - 19.62\,x$

$x = 82$ m

11 Horizontal projection

Shooting horizontally and dropping

- Balance a half-metre rule on the edge of a table and place two small balls as shown in Figure 11.1.
- Twist the rule quickly to drop one ball vertically and knock the other one horizontally off the table at the same time.
- Compare the time at which they reach the ground.

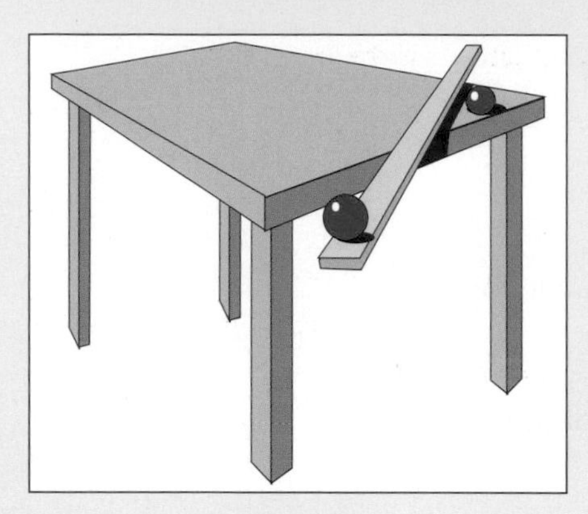

Figure 11.1 Shooting horizontally and dropping

Horizontal projection

The above experiment shows that an object fired horizontally falls vertically with the same acceleration as one which is just dropped. A body that is thrown horizontally from the top of a cliff takes the same time to reach the bottom of the cliff as a body dropped vertically. The cartoon in Figure 11.2 shows this in another way.

Bodies that are thrown (projected) are called **projectiles**. For projectiles, the vertical motion is independent of the horizontal motion. Projectiles will accelerate downwards with the acceleration of 9.8 m s^{-2} caused by gravity,

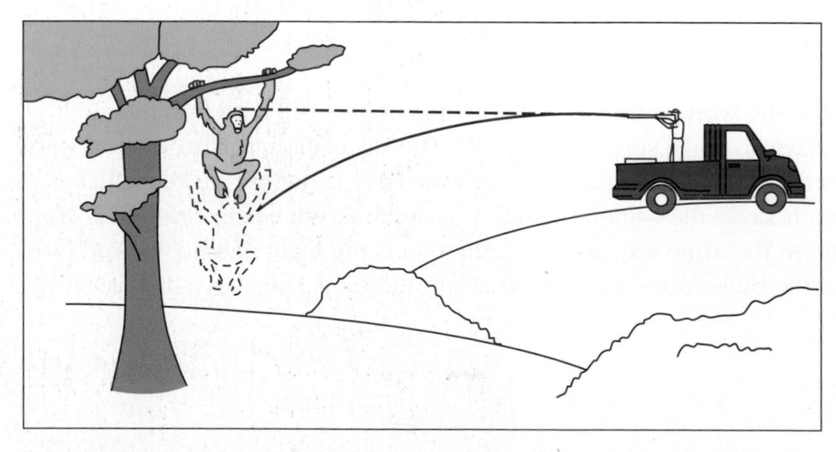

Figure 11.2 The monkey let go at the same time as the hunter fired horizontally at it. It fell the same vertical distance as the bullet and, sadly, met its end!

however fast or slowly they are moving horizontally. The horizontal velocity of the projectile will remain constant (ignoring air resistance).

The constant horizontal velocity and the constant vertical acceleration of a projectile result in a curve path through the air, called a *parabola*.

Measuring the speed of a snooker ball

- Use a cue to fire a snooker ball horizontally off the table and measure how far it goes, both horizontally and vertically (Figure 11.3).
- Calculate the time of flight from the vertical distance.
- From the time of flight and the horizontal distance travelled, calculate the horizontal velocity, assuming that it is constant.

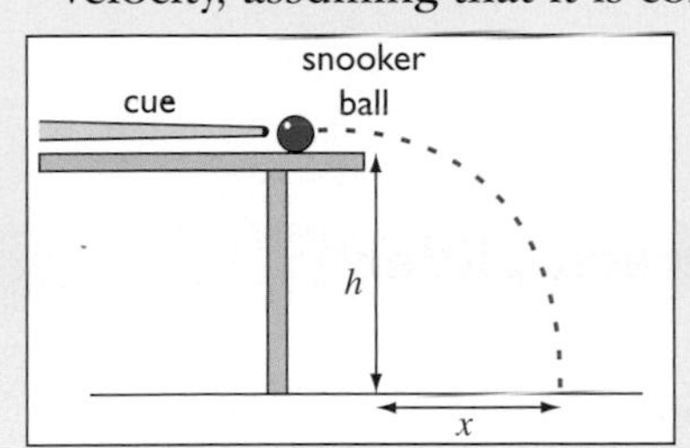

Figure 11.3 Finding the speed of a horizontally projected snooker ball

WORKED EXAMPLE

A tennis ball passes horizontally just over the net and lands just inside the base line of the court. The net has a height of 1.07 m and is 11.9 m from the base line. Find the horizontal speed of the tennis ball. (Ignore air resistance and spin.)

Use vertical motion to find the time in the air. We have

$v = ?$ (irrelevant)

$u = 0$

$x = 1.07\ \text{m}$

$a = 9.8\ \text{m s}^{-2}$

$t = ?$

so we use

$$x = ut + \tfrac{1}{2}at^2$$

$$= \tfrac{1}{2}at^2 \quad \text{(because } u = 0\text{)}$$

$$1.07\ \text{m} = \tfrac{1}{2} \times 9.8\ \text{m s}^{-2} \times t^2$$

$$= 4.9\ \text{m s}^{-2} \times t^2$$

$$t^2 = \frac{1.07\ \text{m}}{4.9\ \text{m s}^{-2}}$$

$$= 0.22\ \text{s}^2$$

$$t = 0.47\ \text{s}$$

Use horizontal motion to find speed (assumed constant):

$$\text{speed} = \frac{\text{distance}}{\text{time}} = \frac{11.9\ \text{m}}{0.47\ \text{s}} = 25.5\ \text{m s}^{-1} \text{ (about 57 mile/h)}$$

12 Newton's first law

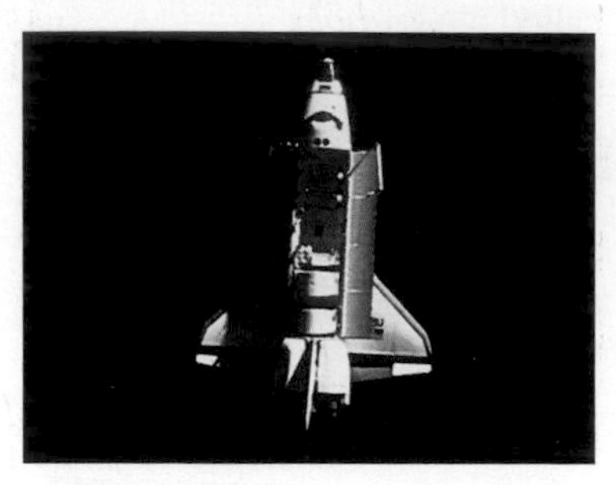

Figure 12.1 Alone in space

Moving freely

Imagine a spacecraft in outer space (Figure 12.1), a long way away from other stars and planets, so that you could ignore the effect of any other bodies. You might say that the spacecraft is floating in space. It is not immersed in a fluid, but it behaves a bit as if it were. The spacecraft can move freely in all directions. It is not affected by other bodies. On Earth, you can imitate some aspects of the free movement of the body in space if you examine movement in one direction only, for instance the horizontal motion of something that can move freely horizontally.

Observing the motion of an air track glider

- Set up the air track so that it is horizontal and place a glider on it (Figure 12.2). Feel how freely the glider can move horizontally.
- Set the glider at rest and then leave it alone. Observe what happens.
- Next give the glider a push to set it moving. Again observe what happens.

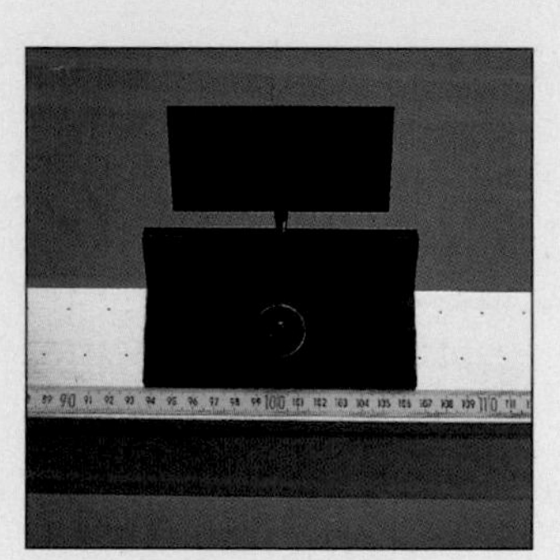

Figure 12.2 A glider on an air track

What causes acceleration?

Within the imperfections of the apparatus, you find that a stationary glider on an air track remains stationary – its velocity remains at zero. And a moving glider keeps going at a constant velocity; it does not speed up or slow down. For both, the acceleration is zero.

If you want a freely moving body to accelerate, you need to push or pull it. When you do so, you are applying a **force** to it. Something has to cause a body to accelerate; that something is a force.

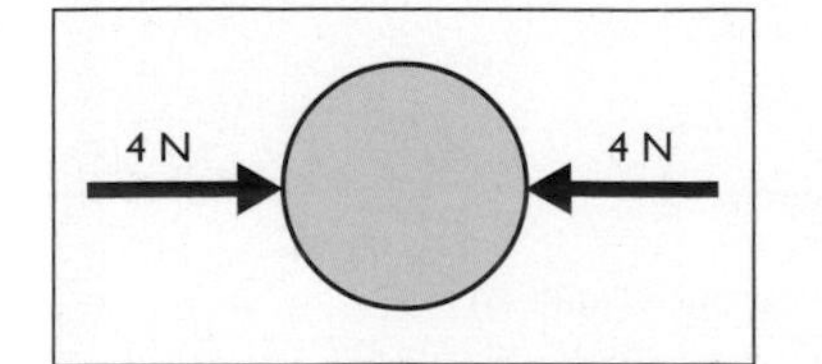

Figure 12.3 Two balanced forces

Newton's first law of motion

Forces are measured in newtons (N). If you apply a pair of equal and opposite forces to the glider, then it will not accelerate. The forces on the body cancel out. They are balanced, as in Figure 12.3.

If you apply more forces and the forces still balance, the body still will not accelerate, as in Figure 12.4.

When there are no forces acting on a body, or when the forces on a body cancel out, the body will not accelerate. The body is said to be in **equilibrium**.

What you need for acceleration is an unbalanced force or a set of forces that have a **resultant force** (or simply, resultant) in some direction, as in Figure 12.5.

Newton's first law of motion summarises these observations:

> A body will remain at rest or continue to move with a constant velocity as long as the forces on it are balanced.

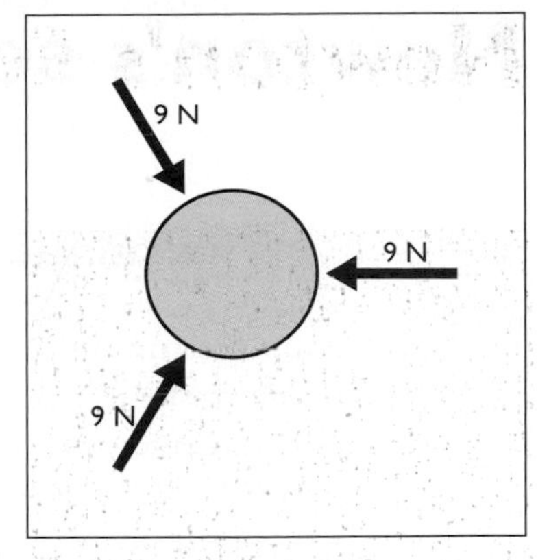

Figure 12.4 Three balanced forces

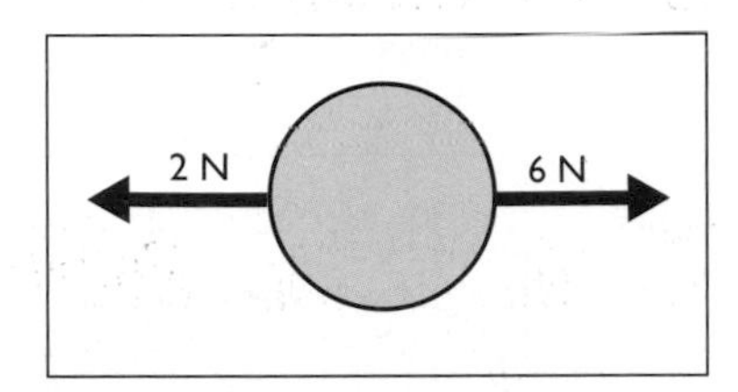

Figure 12.5 Two unbalanced forces: the resultant force is 4N to the right

Inertia

A body behaves as though it is reluctant to change its velocity. In other words, a body will remain at rest or will continue to move with a constant velocity unless it is forced to do otherwise. This reluctance to change velocity is the **inertia** of the body.

It is much harder to change the velocity of a larger mass than a small one. The bigger the mass of a body, the larger its inertia.

Testing inertia

- Set up two tin cans as in Figure 12.6.
- Give each of the stationary tin cans a similar, sharp knock with the side of your hand. Observe the difference.
- Next, pull back both cans to the same height and then release them. Try to stop each can as it reaches the bottom of its swing.

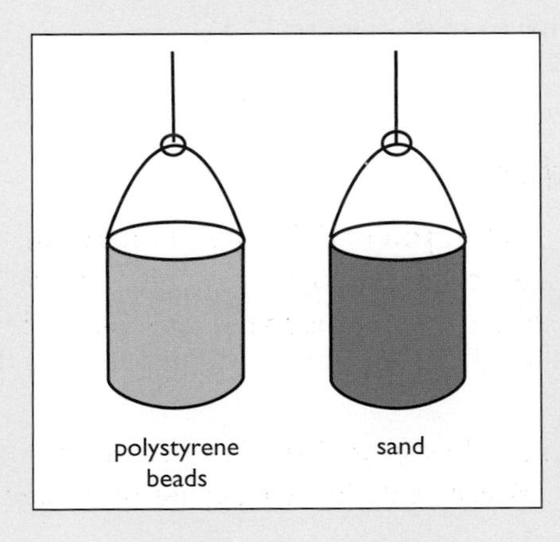

Figure 12.6 Testing inertia with suspended tin cans

13 Newton's second law

The relationship between force, mass and acceleration

- Set up the runway and tilt it just enough so that a trolley runs down it at constant speed when given a push.
- Use a forcemeter to apply a constant force to the trolley and measure the acceleration (Figure 13.1). Repeat for a range of forces and plot a graph of acceleration against force.
- Next, repeat the experiment using a constant force and adding masses to the trolley.
- Measure the acceleration for a range of measured masses. Plot a graph of acceleration against $\frac{1}{\text{mass}}$.

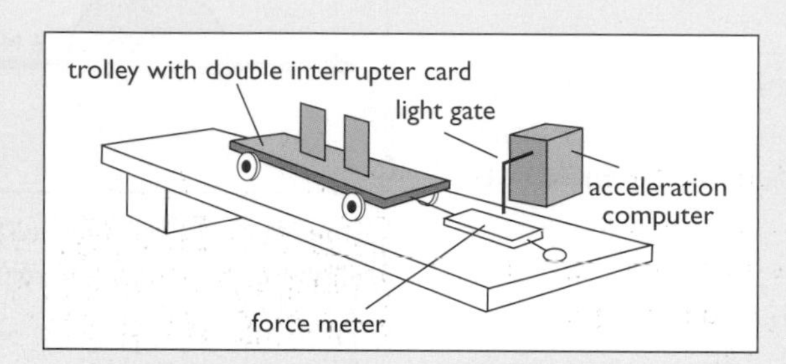

Figure 13.1 Accelerating a trolley

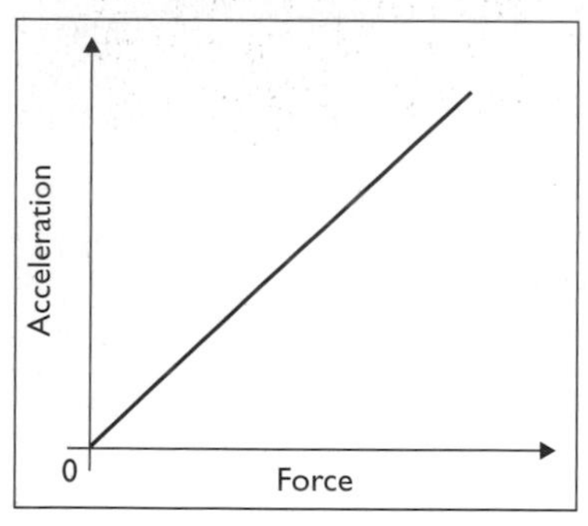

Figure 13.2 An acceleration–force graph for the trolley experiment

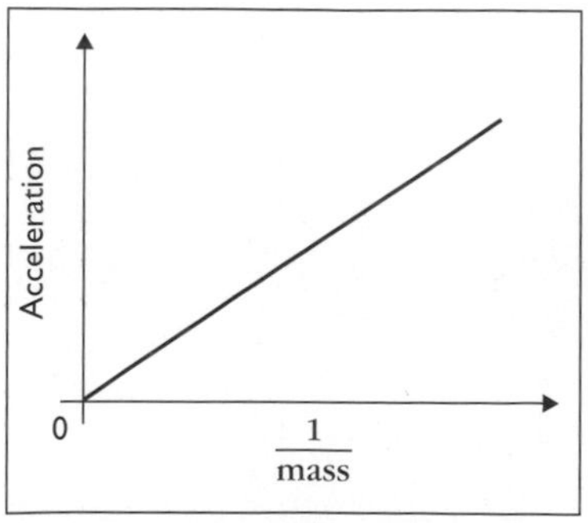

Figure 13.3 A graph of acceleration against 1/mass for the trolley

What determines the acceleration?

If the resultant force acting on a body increases, so does the acceleration; the acceleration of the body is directly proportional to the resultant force (Figure 13.2).

The same resultant force will give a smaller acceleration to a larger mass; acceleration is inversely proportional to mass (Figure 13.3).

These two results can be summarised by

$$\text{force} \propto \text{mass} \times \text{acceleration}$$

or

$$\text{force} = k \times \text{mass} \times \text{acceleration} \quad \text{where } k \text{ is a constant}$$

In the International System, the value of the constant k is equal to 1, so the equation becomes

$$\text{force} = \text{mass} \times \text{acceleration}$$

The definition of the newton

The equation above also defines the newton, since

$$1\ \text{N} = 1\ \text{kg} \times 1\ \text{m s}^{-2}; \qquad \text{N} = \text{kg m s}^{-2}$$

So **one newton** is the resultant force that will give a mass of 1 kg an acceleration of 1 m s^{-2}. Some typical force values are listed in Table 13.1. An example of a force producing an acceleration is shown in Figure 13.4.

Figure 13.4 The Space Shuttle's booster rockets give a force of 24 000 000 N on a mass of 2 000 000 kg. What is the acceleration?

Table 13.1 *Typical force values*

1 N	the gravitational force on an apple
10 N	the gravitational force on 1 kg
100 N	a firm push
600 N	the gravitational force on you
10 000 N	the gravitational force on a car

Newton's second law of motion

It may seem obvious, but it is worth stating that the direction of acceleration is the same as the direction of the resultant force. If the resultant force is downwards, the acceleration is downwards. If the resultant force is west, so is the acceleration. This, combined with the above equation, produces one version of **Newton's second law of motion**, which states that:

> The acceleration of a body is proportional to the resultant force and takes place in the direction of the force.

There is an alternative statement of Newton's second law in Chapter 25.

Newton's laws, like other scientific laws, attempt to describe or predict behaviour, as outlined in Table 13.2.

Table 13.2 *Laws*

Moral laws and laws of the land tell people what they should do.
By contrast, physical laws, like Newton's laws, are just descriptions of the universe as we find it. They don't tell the universe how to behave.

Worked example

A tractor pulls a tree stump of mass 2000 kg along rough horizontal ground. The tractor exerts a horizontal force of 1300 N on the tree stump which accelerates at 0.05 m s^{-2}. What is the frictional force between the tree stump and the ground?

Resultant force $F = ma$ so

$F = 2000\ \text{kg} \times 0.05\ \text{m s}^{-2} = 100\ \text{N}$

Since resultant force = pull of tractor − friction

100 N = 1300 N − friction

friction = 1200 N

14 Newton's third law

Figure 14.1 A spacecraft alone in outer space

A body with no forces

Think again about a spacecraft in outer space, a long way from other bodies, so you could imagine that there were no forces acting on the spacecraft (Figure 14.1).

You know from Newton's first law that, if the spacecraft were stationary, it would stay stationary. If it were moving in a straight line, it would keep on moving at the same velocity. Now imagine that you were in space too, just beside the spacecraft and stationary relative to it. There are no forces on you or the spacecraft; you stay stationary relative to it (Figure 14.2).

Figure 14.2 You and the spacecraft exert no force on each other, so both are in equilibrium

A single force?

Now think what happens when you apply a single force to the spacecraft by pushing it (Figure 14.3).

Both you and the spacecraft are a long way from other bodies. So you would be standing on nothing. You can think about what would happen if it did occur. This is called 'doing a thought experiment'. An experiment in the laboratory helps to understand what might happen.

Figure 14.3 What happens when you push the spacecraft?

Pushing the body

- **Find a large body that is on very smooth wheels.**
- **Stand on free-moving rollers, or a good skateboard, to imitate something of the free movement of deep space.**
- **Observe what happens when you push the body (Figure 14.4).**

Figure 14.4 What happens when you push now you are on wheels?

Pairs of forces

When you push an object to the left, it accelerates to the left. But that isn't the end of the story; the body pushes you, but to the right. It helps to draw separate diagrams of both you and whatever you are pushing to show what is happening to each (Figure 14.5).

You started off thinking about a single force and a very simple situation in which that force might occur. But when you apply a single force to anything, that body applies another force to you.

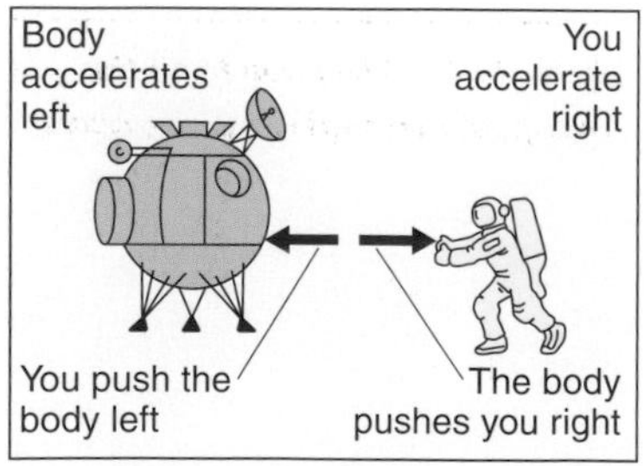

Figure 14.5 The body accelerates to the left; you accelerate to the right

Forces occur only in pairs

The body accelerates to the left because you push it to the left. And you accelerate to the right because it pushes you to the right.

The fact is that you can't have one force on its own. When any one body exerts a force on a second body, the second body exerts a force on the first body. One force cannot occur without the other.

Who is pushing whom?

- Stand with a friend, each of you on a skateboard, and pull a rope connecting you. Observe what happens.
- Then hold the rope and get your friend to do the pulling. Observe again.
- Then both pull.
- Repeat the experiments using spring forcemeters to measure the forces you each exert (Figure 14.6). Record the measurements.

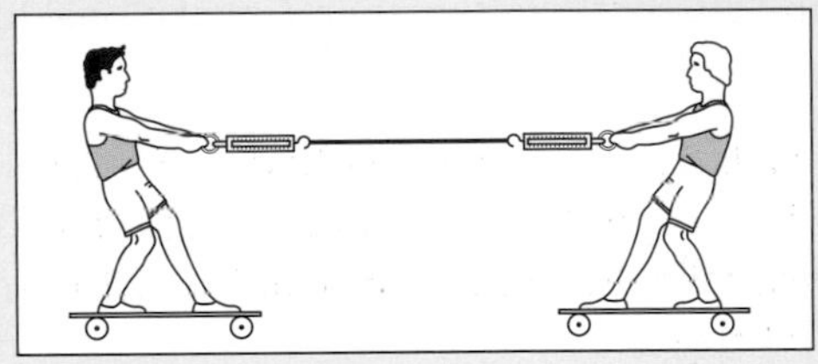

Figure 14.6 Measuring the force you exert

Newton's third law of motion

Look at Figure 14.7. When you burn the string holding the two spring-loaded trolleys together, they accelerate apart. If you measure their masses and their accelerations, you find that the forces they exert on each other are equal. It doesn't matter which trolley has the spring, or whether both do. While trolley A pushes B, trolley B pushes A with an equal force in the opposite direction.

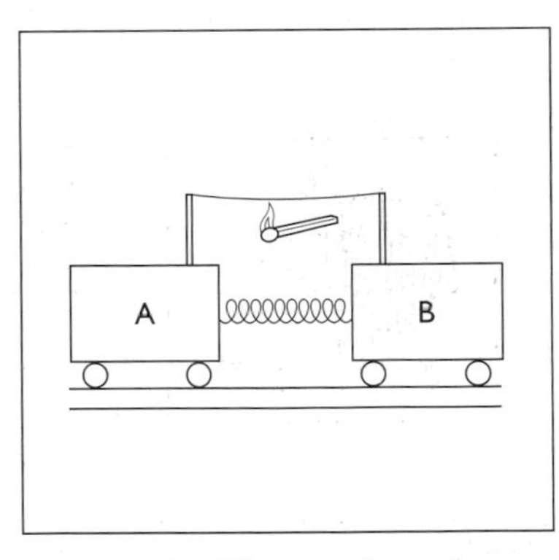

Figure 14.7 When you burn the string, the two trolleys spring apart

The important principle behind these observations led Newton to formulate his third law. As you would expect by now, a single force on its own cannot exist. There are always two forces and **Newton's third law of motion** says about these forces:

> While body A exerts a force on body B, body B exerts an equal and opposite force on body A.

Perhaps the most important and difficult point to appreciate about this pair of forces is that they act on different bodies. The third-law pair of forces between bodies A and B consists of a single force on body A, and a single equal and opposite force on body B. Each of these forces has its separate effect on a separate body. The two forces do not cancel each other.

There is more about these two forces in the next two chapters, and a complete statement of Newton's third law in Chapter 16.

15 Types of forces

Figure 15.1 There is a gravitational force between the spacecraft and the Earth

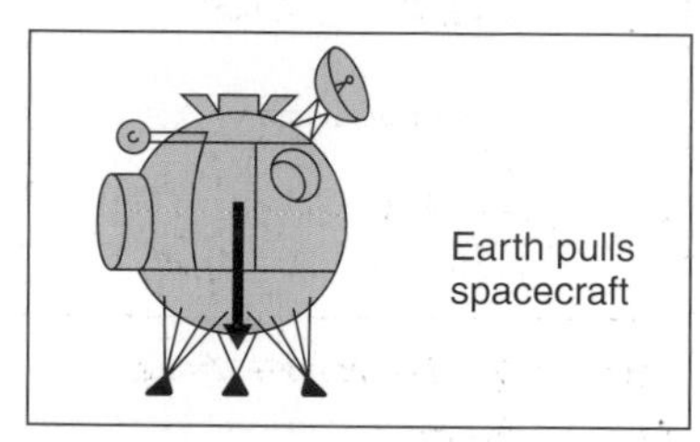

Figure 15.2 The Earth pulls the spacecraft down, so the spacecraft accelerates downwards

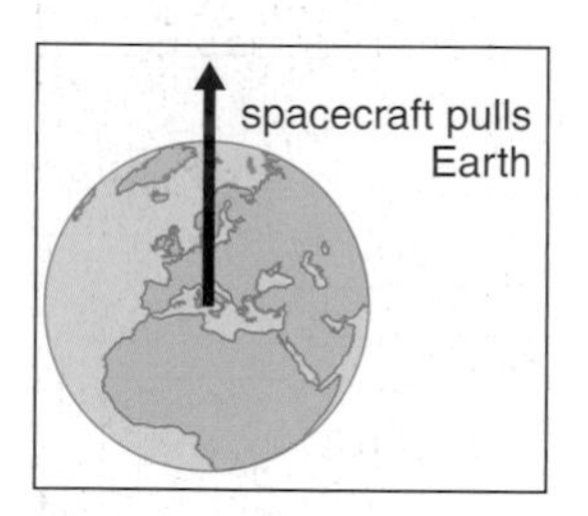

Figure 15.3 The spacecraft pulls the Earth up, so the Earth accelerates upwards

Forces acting at a distance

The forces that bodies in space are most likely to experience are gravitational. Imagine a spacecraft near to a planet, as in Figure 15.1. The planet pulls the spacecraft towards the planet (Figure 15.2). And, just as you would expect, the spacecraft pulls the planet with an equal force towards the spacecraft (Figure 15.3).

The force on the spacecraft is equal and opposite to the force on the planet, and the forces have the same line of action and act for the same time.

The two bodies are exerting forces on each other without being in contact. These are gravitational forces; they act at a distance. Gravitational forces occur between all bodies of all sizes and attract them together. Gravitational forces are one of the four different types of force that exist.

Gravity and gravitational fields

Masses around any body experience gravitational forces. The region in which they experience this force is the body's gravitational field. The gravitational force that a planet applies to a body is called its **weight**. When you are standing on the Earth, the Earth pulls you towards it and provides your weight. You pull the Earth with an equal and opposite force.

You can calculate the weight of a body by using the formula

weight = mass × gravitational field strength

$$W = mg$$

Near the surface of the Earth, the gravitational field strength is about 9.8 N kg^{-1}. You will learn more about this in Unit 5.

Investigating electrostatic forces

- Rub a polythene rod with a duster and then suspend it from a thread.
- Rub another rod and suspend it, too.
- Bring them together (Figure 15.4) and observe what happens.
- Draw diagrams to show the forces on each rod.
- Repeat with two acetate rods, then with one polythene and one acetate rod.

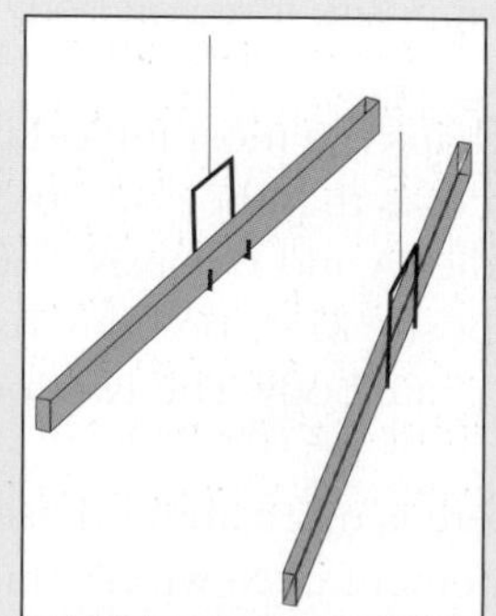

Figure 15.4 Investigating electrostatic forces

Electrostatic forces

When you rub a polythene rod with a duster, the rod becomes charged negatively. Two negatively charged rods repel, as do two positively charged rods; but a positive and a negative charge attract (Figure 15.5). These forces are electrostatic forces. Like gravitational forces, electrostatic forces act at a distance.

Investigating magnetic forces

- **Bring the north-seeking ends of two magnets together.**
- **Observe what happens and then draw diagrams to show the forces on each end.**
- **Repeat with two south-seeking ends and then with one north-seeking and one south-seeking.**

Electromagnetic forces

Between two magnets, or between a magnet and another piece of magnetic material, there are magnetic forces. These forces, too, act at a distance. The forces between static charges and between magnets are related. One of Einstein's (Figure 15.6) great pieces of work was to suggest that electrostatic forces and magnetic forces are just the same type of force viewed from relatively different positions. His theory of special relativity made it possible to understand them as a single force, called the electromagnetic force.

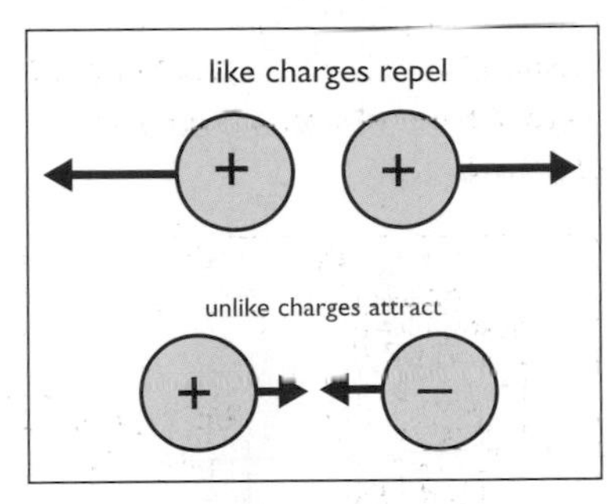

Figure 15.5 Electrostatic forces between charges

Nuclear forces

There are forces that act inside the nucleus of an atom which are neither gravitational nor electromagnetic. These nuclear forces are important – they hold the nucleus together. Electrostatic repulsion alone would cause the positively charged protons in the nucleus to fly apart. But there is a force, the strong nuclear force, that binds them together.

The weak nuclear force is involved, for example, in beta decay when a neutron decays to a proton, with the emission of an electron. The nuclear forces act over very small distances and are often referred to as short-range forces.

Forces are often described as interactions, particularly when small particles are involved. This helps to emphasise that they always involve the effect of one body acting on another.

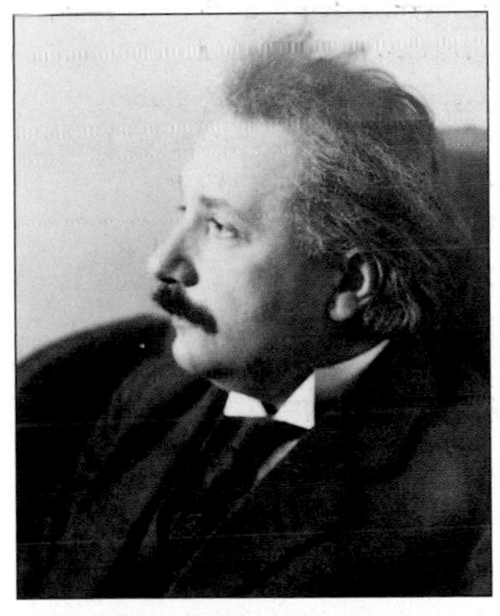

Figure 15.6 Einstein's success in providing a unified theory of electricity and magnetism encouraged him and others in the search for a theory, a unified field theory, that would describe all the different types of force

Contact forces

When you push a car, pick up a book or push a cork into a bottle, the forces are not obviously gravitational, electromagnetic or nuclear. These are called **contact forces**, because the bodies between which they exist are very close together. Contact forces are, in fact, electrostatic forces acting over very short distances. They are caused by forces between the outer layers of electrons of the two bodies in contact. When you push something, the electrons on the outside of your hand repel the electrons on the outside of the body you are pushing. Frictional forces, forces on bodies moving through liquids, and air resistance are all contact forces.

16 Free-body force diagrams – I

Figure 16.1 Situation diagram for a man near the Earth

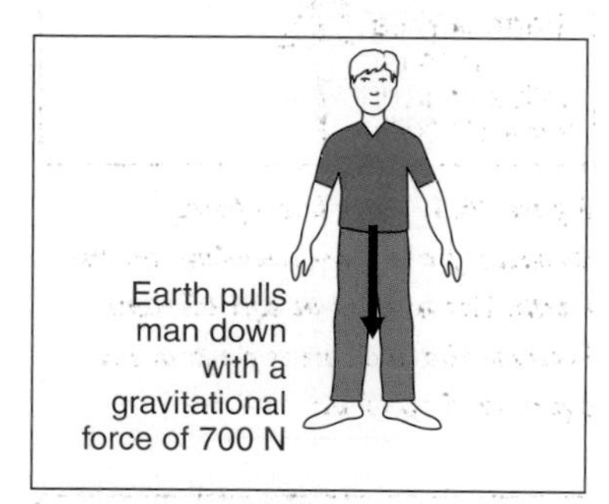

Figure 16.2 Free-body force diagram for a man near the Earth

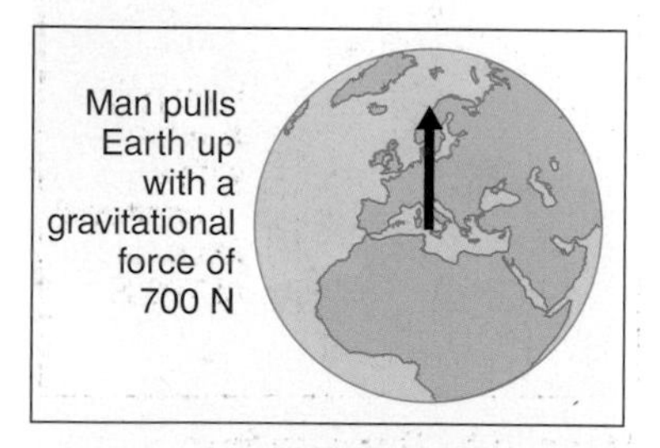

Figure 16.3 Free-body force diagram for the Earth near a man

Situation diagrams and free-body force diagrams

You know that forces always occur in pairs, one force acting on one body, and one force on another body. If you draw both these forces in the same diagram it is confusing because, in all but the simplest situations, it is unclear which forces act on which body. Free-body force diagrams are a neat and effective way to show the forces on a body; they are diagrams of a *single* body, showing the forces on that body *only*.

Figure 16.1 shows a situation diagram for a man near the Earth. Figures 16.2 and 16.3 show two *free-body force diagrams*, one for the man and one for the Earth. Each free-body force diagram shows a single body, and the forces on that body alone. Free-body force diagrams may look like situation diagrams, but there is one really important difference. In the man's free-body force diagram, he is on his own, without the Earth. That does not mean that the Earth doesn't exist, just that the diagram is only about what is happening to the man.

If you need to draw a free-body force diagram, it sometimes helps first to sketch a rough situation diagram with all the bodies, and then to draw a separate free-body force diagram for each body.

Describing forces

The labels on the forces in the free-body force diagrams for the man and the Earth describe the forces fully. These descriptions consist of five parts:

The Earth pulls the man down with a gravitational force of 700 N

The description states

- what is exerting the force
- on what the force is exerted
- the direction of the force
- the type of the force
- the size of the force.

Similarly,

The man pulls the Earth up with a gravitational force of 700 N

Descriptions like this help identify **Newton III pairs of forces**.

Identifying Newton III pairs of forces

From this work on free-body force diagrams, you should be able to see that the forces in a Newton III pair of forces have the same line of action – if you extend the force lines on the diagram of the situation, then they will pass through each other. The forces also act for the same time. In addition, you should realise that the two forces of a Newton III pair have to be the same type of force. If A pulls on B with a gravitational force, then B pulls on A with a gravitational force also. This information allows a complete statement of Newton's third law:

> While a body A exerts a force on a body B, body B exerts a force on body A. The forces are equal, opposite and of the same type; they have the same line of action and act for the same time.

If you have described one force fully, then you can identify the Newton III pair to this force from its description. Simply exchange the two bodies in the statement and reverse the direction of the force. For example:

> The Earth pulls the man down with a gravitational force of 700 N

> The man pulls the Earth up with a gravitational force of 700 N

So if

> Freda pushes Jo left with a contact force of 35 N

> Jo pushes Freda right with a contact force of 35 N

Figure 16.4 Situation diagram for a man standing on the Earth

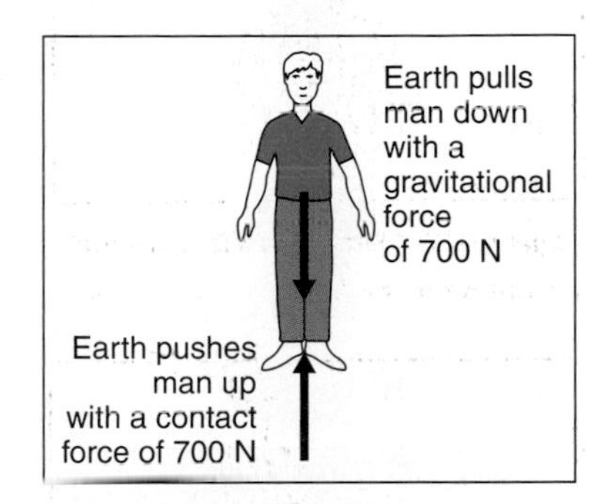

Figure 16.5 Free-body force diagram for a man standing on the Earth. The man is in equilibrium because the two forces on him are equal and opposite

Standing on a planet

Two bodies often exert more than one pair of forces on each other. Think about a body resting on the surface of a planet. Figure 16.4 shows the diagram of this situation for a man on the Earth, and Figures 16.5 and 16.6 show the two free-body force diagrams for the man and the Earth.

As before, there is a pair of gravitational forces between the man and the Earth, one on the man and one on the Earth. But there are also contact forces, again one on the man and one on the Earth.

First think about the forces on the man. The Earth pulls the man down with a gravitational force. The Earth also pushes the man up with a contact force on his feet.

Now think about the forces on the Earth. The man pulls the Earth up with a gravitational force, and also pushes the Earth down with the contact force of his feet.

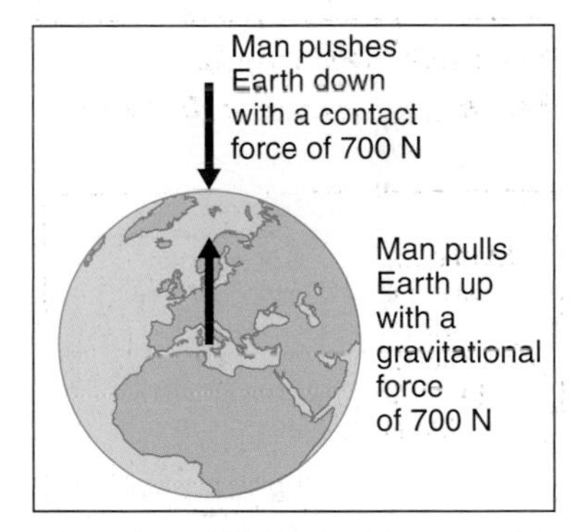

Figure 16.6 Free-body force diagram for the Earth. The Earth is in equilibrium because the two forces on it are equal and opposite

17 Free-body force diagrams – 2

More than two bodies

If you draw free-body force diagrams for all the bodies in a complicated situation, it can help you to understand how the forces on the bodies relate to each other.

Worked example

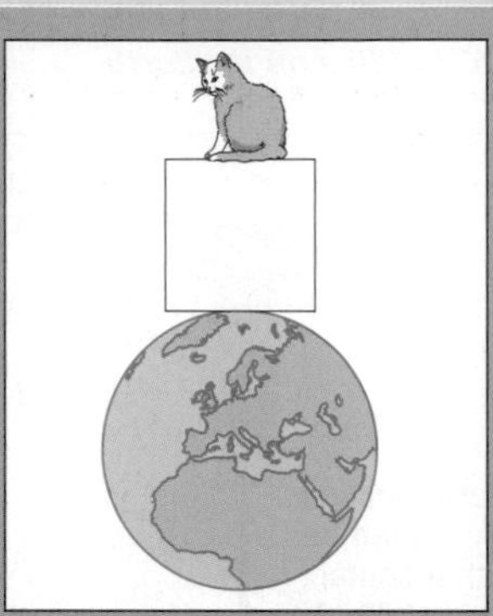

Figure 17.1 shows a cat sitting on a box. Assume that the cat, the box and the Earth are in equilibrium. First draw the three bodies separately to start a free-body force diagram for each.

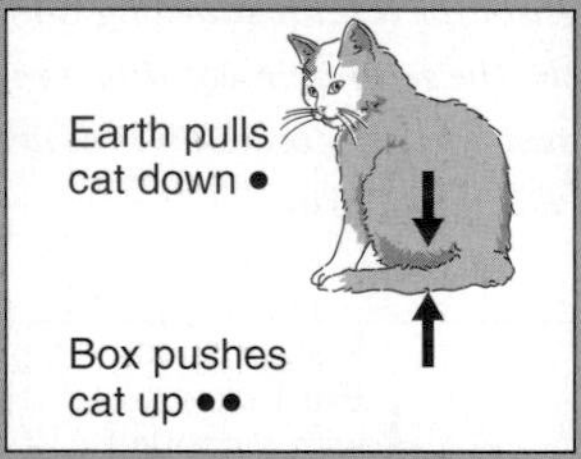

Figure 17.1 The cat, the box and the Earth

Earth pulls cat down •

Box pushes cat up ••

Figure 17.2 Free-body force diagram for the cat

Forces on the cat

Look carefully at Figure 17.2. The third word in both of the force descriptions is 'cat'. The Earth pulls the cat down, and the box pushes the cat up. Grammatically, the object of both of these sentences is the cat. This is because both of these forces are acting on the cat. (If you describe forces simply like this, it helps you to check that you have included only forces that act on that body on a body's diagram.) These two forces are the only significant forces on the cat, and, since the cat is in equilibrium, the forces must be equal and opposite, in agreement with Newton I.

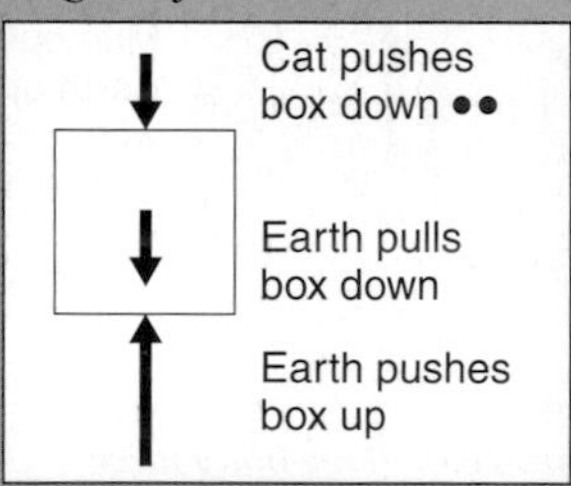

Figure 17.3 Free-body force diagram for the box

Forces on the box

There are two downward forces on the box (Figure 17.3): the Earth pulls it down, and the cat pushes it down. The Earth itself pushes the box up. Now the third word in each force description is 'box'. The box is in equilibrium, so, as a consequence of Newton I, the upwards force must be equal to the sum of the downwards forces.

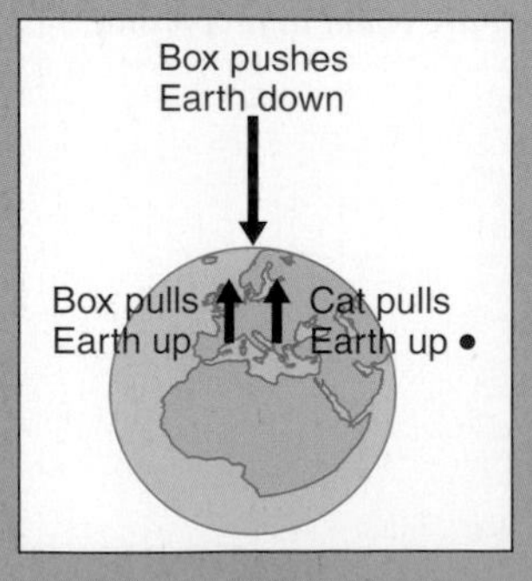

Figure 17.4 Free-body force diagram for the Earth

Forces on the Earth

In Figure 17.4, you can see the forces on the Earth. Both the box and the cat pull the Earth up, and the box pushes the Earth down. The Earth is in equilibrium, so the sum of the forces upwards must be equal to the downwards force.

There are, of course, gravitational forces between the cat and the box, but these are so small that you can ignore them. The single and double bullets show two Newton III force pairs. Work out the others.

Differences between Newton I and Newton III

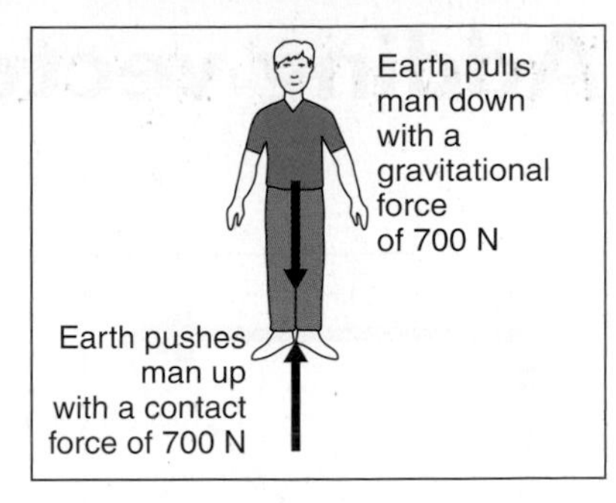

Figure 17.5 These two forces are equal and opposite, but they are not a Newton III pair

Figure 17.5 again shows the free-body force diagram for a man standing on the Earth. The man is in equilibrium. From this you can deduce that the forces on the man are balanced: the upward force on him is equal to the downward force. But these forces are *not* a Newton III pair. There are a number of reasons why we know that.

First, the upward force from the Earth and the downward force from the Earth both act on the same body, the man. With a Newton III pair, the forces act on different bodies.

Secondly, the upward force on the man is a contact force and the downward force is gravitational. With a Newton III pair, the forces are of the same type.

Thirdly, these two forces are only equal when the man is in equilibrium, and he need not necessarily be so. As you know, if the man is above the Earth as in Figure 16.1, there is no contact force but there is still a gravitational force. With a Newton III pair the forces are *always* equal and opposite, and one force cannot occur without the other.

The man in Figure 16.5 just happens to be in equilibrium under the action of two forces and so, in accordance with Newton I, these forces are equal and opposite.

These differences are summarised in Table 17.1.

Table 17.1 *Some important differences between Newton I and Newton III*

Newton I	Newton III
A law about the forces *on a single body*	A law about a pair of forces *on two different bodies*
Concerns *any number of forces*	Always concerns *two forces only*
The forces can be *different types*	The forces are always the same type
If there are two forces *and* the body is in equilibrium, the forces are equal and opposite	The two forces are *always* equal and opposite
Newton I *only* applies when a body is in equilibrium	Newton III *always* applies

18 Adding vectors

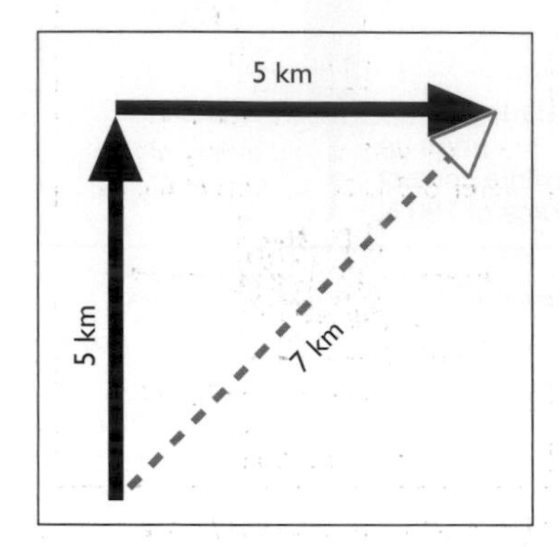

Figure 18.1 The resultant of these two displacements is just over 7 km north-east

Adding scalars and vectors

Forces have size and direction. They are vectors like displacement and velocity. There are special rules for adding vectors. You can add scalars by simple arithmetic. A distance of 5 km followed by another distance of 5 km gives a total distance of 10 km. But when adding vectors like displacement, you must take into account the direction. Remember that 4 km to work and then 4 km back home gives a total displacement of zero rather than 8 km.

If you have a displacement of 5 km north and then a displacement of 5 km east, you end up north-east of your starting point, as Figure 18.1 shows. You can find the final displacement (called the *resultant* displacement) by drawing a diagram to represent the displacements in turn. If you draw the diagram to scale, you can take measurements from the diagram to find the final displacement. Or, you can draw a sketch diagram and calculate the resultant mathematically. In this case the resultant displacement is just over 7 km north-east of the starting point. You can add any vectors like this, representing them in size and direction by lines on a diagram and then calculating their resultant or measuring it from the diagram.

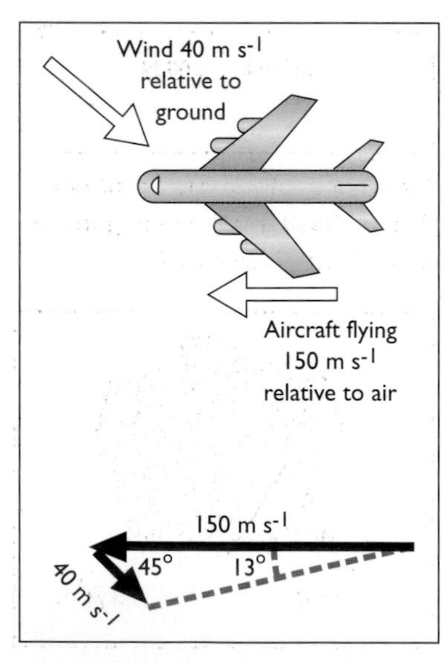

Figure 18.2 The velocity of the aircraft relative to the ground is the sum of its velocity relative to the air, plus the velocity of the air relative to the ground

Adding velocities

The instruments on an aircraft indicate the velocity of the aircraft relative to the air it flies through. But the air through which it flies moves relative to the ground. If the pilots wish to know the aircraft's velocity relative to the ground, they, or their computers, need to add the velocity of the aircraft relative to the air to the velocity of the air relative to the ground.

Figure 18.2 shows an aircraft flying west, at 150 m s^{-1} relative to the air, at the same time as a wind is blowing at 40 m s^{-1} from the north-west relative to the ground. You add the two velocities in just the same way as displacements. The resultant velocity is 125 m s^{-1} at an angle of 13° south of west.

Adding forces

Forces are vector quantities. If you wish to add forces together to find their resultant, you need to take account of their directions.

In Figure 18.3, Freda and Jo are pulling in opposite directions with forces of 600 N and 700 N. As the force diagram shows, it is easy to see that the resultant is 100 N in the direction that Jo is pulling. If they push in the same direction (Figure 18.4), the resultant is 1300 N. When Freda and Jo push at right angles (Figure 18.5), you need to use a vector diagram to find the resultant force, as follows.

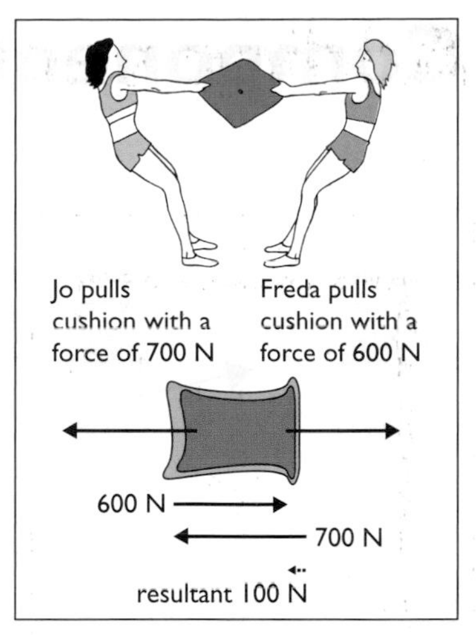

Figure 18.3 Freda and Jo are pulling a cushion. What is the resultant force on it if Jo pulls with a force of 700 N and Freda with 600 N?

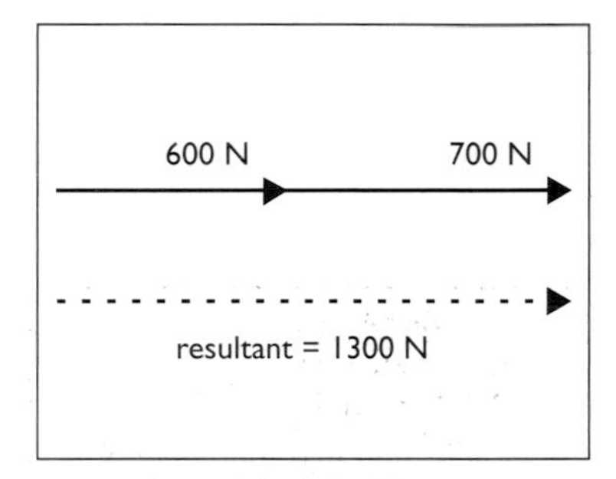

Figure 18.4 What is the resultant force of Freda and Jo pushing in the same direction on a rock?

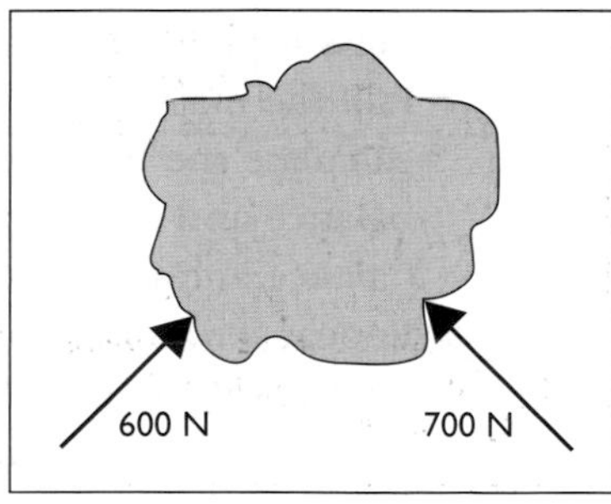

Figure 18.5 What is the resultant now Freda and Jo push the rock at right angles?

Worked example

Represent the vectors you wish to add, in size and direction, with lines on a vector diagram. Draw the lines one after another, starting each line where the one before ends (Figure 18.6). The resultant is the line from where you start to where you end.

You can work out the resultant either by drawing the vectors to scale, or by using trigonometry. If you use trigonometry, draw a diagram roughly to scale first, so that you can check if the resultant is roughly correct.

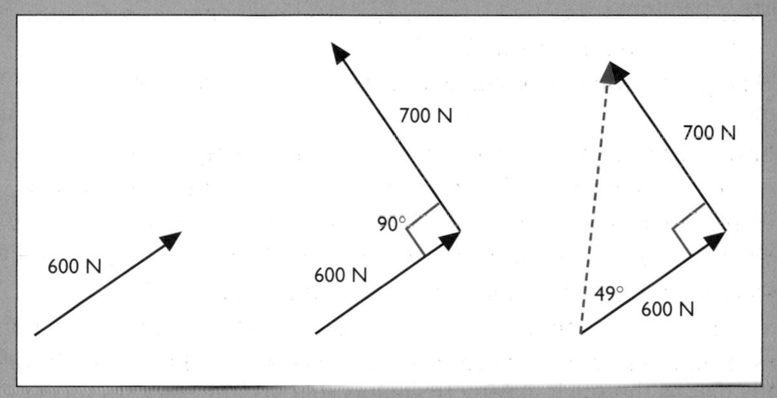

Figure 18.6 Drawing the resultant for Figure 18.5: (a) first draw one of the vectors; (b) then draw the other, at 90°, from the end of the first one; (c) the resultant is the line that completes the triangle

Drawing vector force diagrams

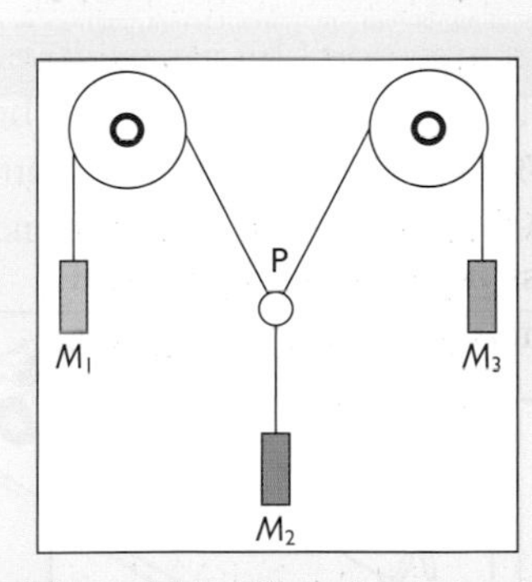

Figure 18.7 Three forces in equilibrium

- Set up the pulleys and mass hangers as shown in Figure 18.7.
- Hang masses on the three mass hangers and let them move until stable. Draw the string pattern on the paper behind the pulleys.
- Determine the forces on point P by measuring the masses.
- Draw a vector diagram to find the resultant of the two upward forces. Check that this is equal and opposite to the downward force.

20 Fluid forces

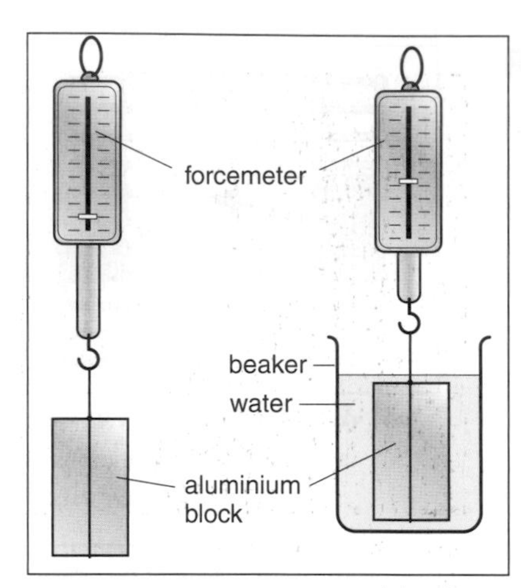

Figure 20.1 The block needs less force from the force meter to support it when in water

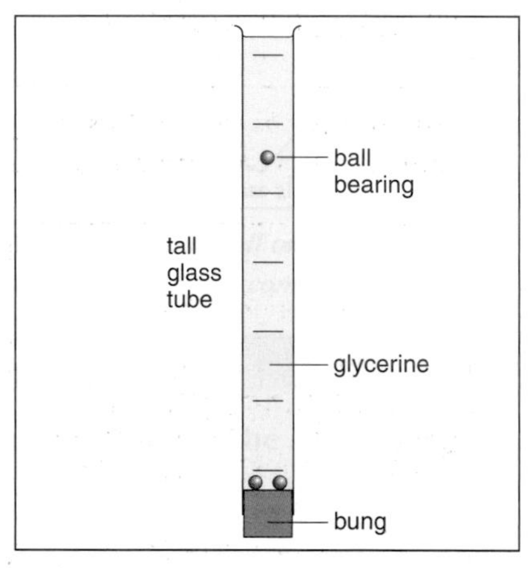

Figure 20.2 Ball bearings fall slowly through glycerine

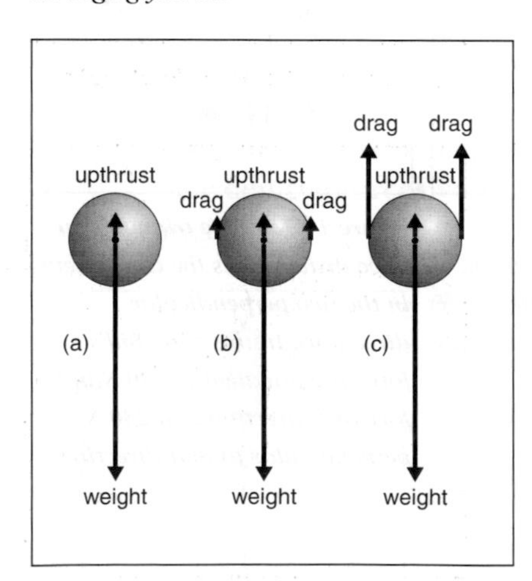

Figure 20.3 When the ball bearing is in equilibrium, its velocity is constant

Upthrust

Fluids exert pressure on immersed objects. The bottom of an immersed object experiences greater fluid pressure than the top, because it is at a greater depth. The pressure difference between top and bottom results in an upward force, called **upthrust**, acting on the immersed object. An object appears to weigh less when it is immersed in a fluid (Figure 20.1).

Drag and viscous forces

A ball bearing in Figure 20.2 accelerates when released, but soon reaches a constant speed known as its **terminal speed** (or sometimes **terminal velocity**). At this stage in its motion, the ball bearing is no longer accelerating; it is in *equilibrium*. The resultant of the forces acting on it must be zero.

At the instant of release there are two forces acting on the ball bearing – its weight and the upthrust from the glycerine. The weight is greater, giving a resultant downward force. Both weight and upthrust remain unchanged throughout the fall. If the ball eventually reaches a constant speed and is in equilibrium, there must be a third force in an upward direction. This is the **drag**. Drag is a force exerted by a fluid which resists the movement of an object through that fluid.

If you try to move your arm through water, you will notice the drag, as you will if you try to move a large balloon through air.

Drag increases with the speed of the object. The faster an object is moving, the greater the drag acting against its motion. The three free-body force diagrams in Figure 20.3 show the ball bearing (a) stationary at the instant of release, (b) accelerating at a reduced rate and (c) moving at its terminal speed.

At the terminal speed,

weight = upthrust + drag

Falling through air

If you drop a ball bearing through air rather than through glycerine, it needs to travel much further to reach a terminal speed. And the terminal speed would be much higher. Air is less viscous and less dense than glycerine. It produces less drag for the same speed and less upthrust.

Skydivers in 'free-fall' reach terminal speeds that can exceed 100 mph (Figure 20.4). Since the atmosphere is less dense at higher altitude, they have a greater terminal speed at a height of 3000 m than they do at 1000 m.

Figure 20.4 Most of the time, skydivers are in equilibrium – their velocity is constant

Figure 20.5 The paper rises as you blow over it

Aerodynamic lift

Blow steadily over the top of a sheet of paper as shown in Figure 20.5. The end of the paper moves upwards. The pressure of the moving air over the top of a sheet of paper is less than the pressure of the stationary air underneath. The pressure difference creates an upward resultant force on the sheet of paper.

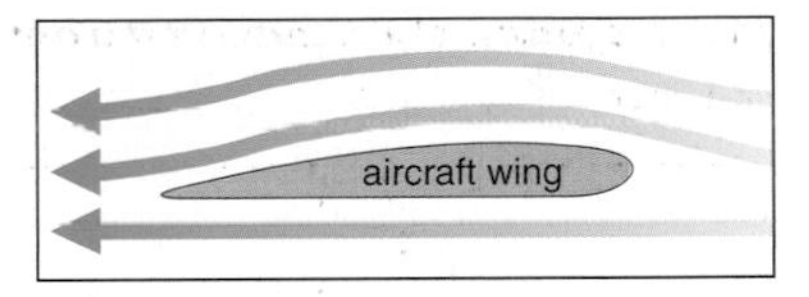

Figure 20.6 Air has to travel faster across the top of an aircraft wing

Air moving across the top of an aircraft wing (Figure 20.6) has to travel further than that moving along the bottom. It travels faster across the top, so the air pressure here is lower. The force resulting from this pressure difference produces the **aerodynamic lift** required for an aircraft to fly.

Bodies moving at constant velocity

You know that, when a body is moving with constant velocity, it is in equilibrium. The forces on it are balanced.

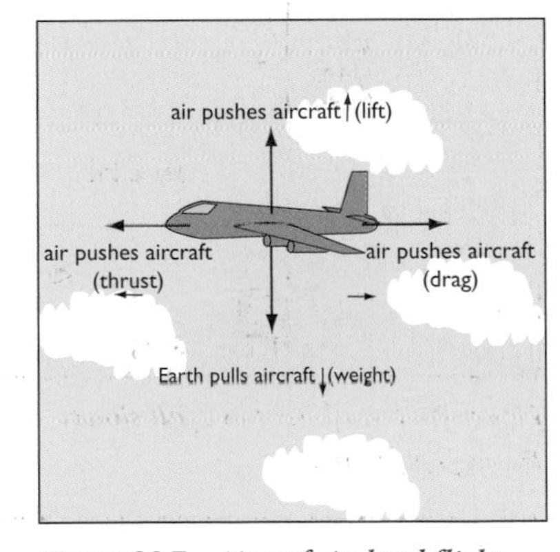

Figure 20.7 Aircraft in level flight

- *Aircraft in level flight:* The air surrounding an aircraft produces three contact forces on it (Figure 20.7). The aircraft's jets push air backwards, and the Newton III pair to this is the force of the air pushing the aircraft forwards. This is the thrust. As air travels over the upper and lower surfaces of the wing, it produces an upwards force called lift. The air produces drag as the aircraft travels through it. If the aircraft is travelling at constant velocity, the forces on it are balanced. Then, the drag is equal to the thrust, and the lift is equal to the weight.

- *Aircraft climbing:* If an aircraft is travelling at constant velocity, the forces on it are balanced, whether the aircraft is climbing (Figure 20.8) or in level flight. The upwards components of the thrust and lift are equal to the weight plus the downward component of the drag. The forward thrust is equal to the backward drag plus the backward component of the weight.

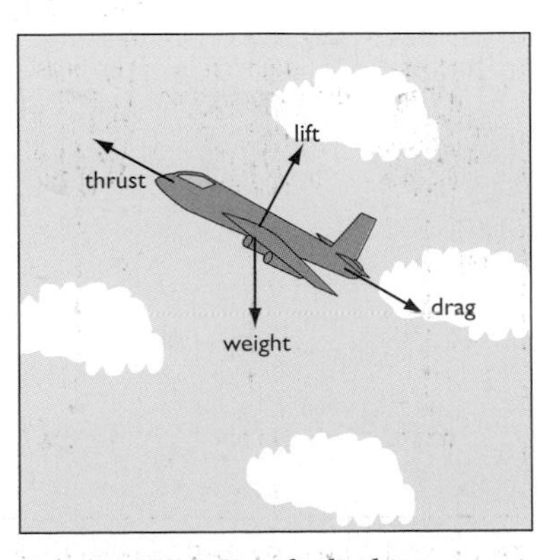

Figure 20.8 Aircraft climbing

21 Free-body force diagrams – 3

Perpendicular and tangential contact forces

As well as perpendicular contact forces, surfaces can produce forces tangential to the surface (along the surface). These **tangential forces** are due to friction.

Cat on a sloping table

Figures 21.1 to 21.3 show a cat on a sloping table.

Figure 21.1 Here is a cat on a sloping table

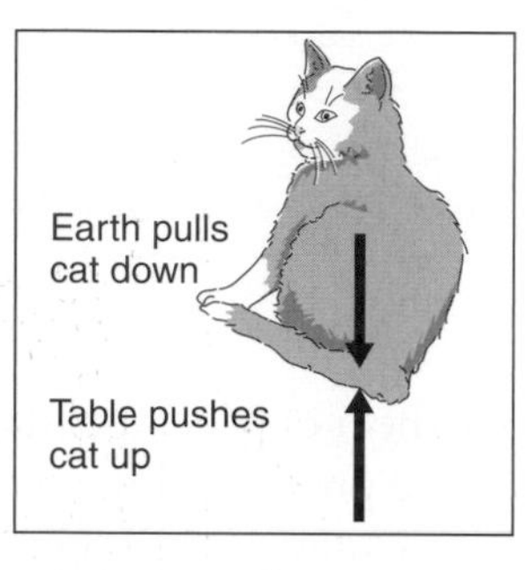

Figure 21.2 The cat remains in equilibrium because the table provides an upwards force on the cat that is equal to the gravitational force on the cat.

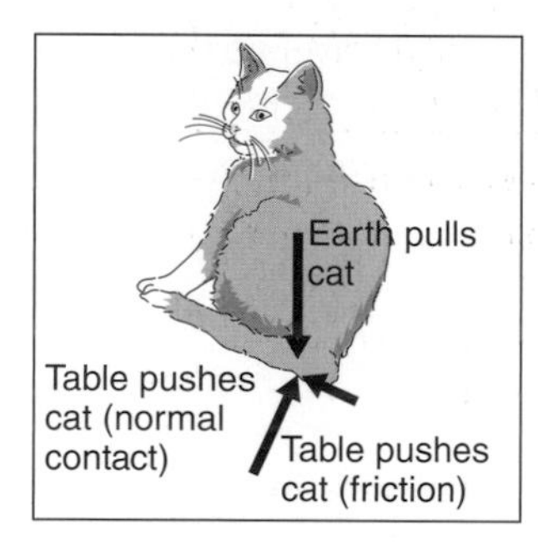

Figure 21.3 You can think of the single upwards force on the cat as being made up of two components: a perpendicular contact force and a tangential contact force of friction. These two components together are equal and opposite to the downwards gravitational force.

Forces on a car

The wheels on a car push the road backwards. So the road pushes the car forwards. It is the force of the road on the wheels that accelerates the car forwards, or keeps it going against the air resistance.

With a four-wheel-drive car (Figure 21.4), all the wheels push the road backwards, so the road pushes all the wheels forwards.

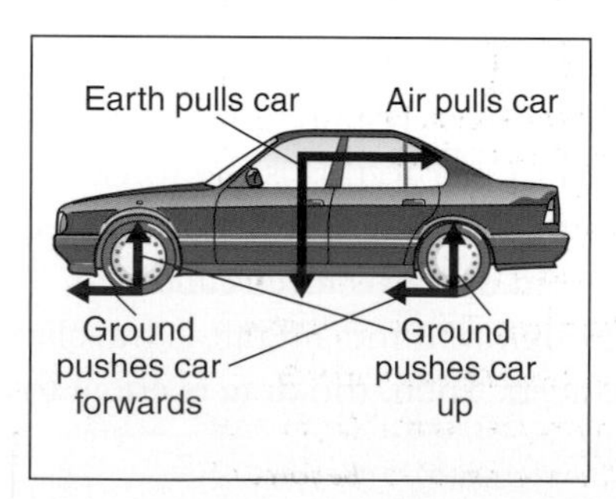

Figure 21.4 Four-wheel-drive car: All the car's wheels push the ground backwards. The Newton III pairs to these forces are the forces of the ground pushing the car forwards. These frictional contact forces are forces that push the car forwards.

With a front-wheel-drive car (Figure 21.5), the road pushes the two front wheels forwards; it also pulls the two rear wheels backwards which results in unhelpful resistance to the car moving forwards.

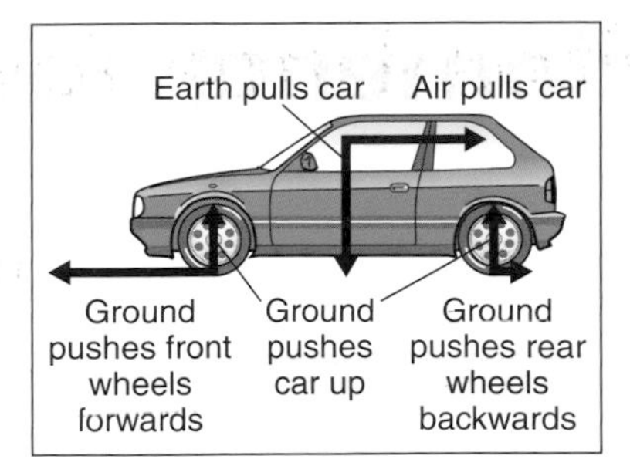

Figure 21.5 Front-wheel-drive car: In this case, only the front wheels push the ground backwards, so the ground pushes these two wheels forwards. The rear wheels are not driven; the ground pushes them backwards.

Horse pulling cart

Figure 21.6 The horse pulls the cart at a steady speed along the surface of the Earth

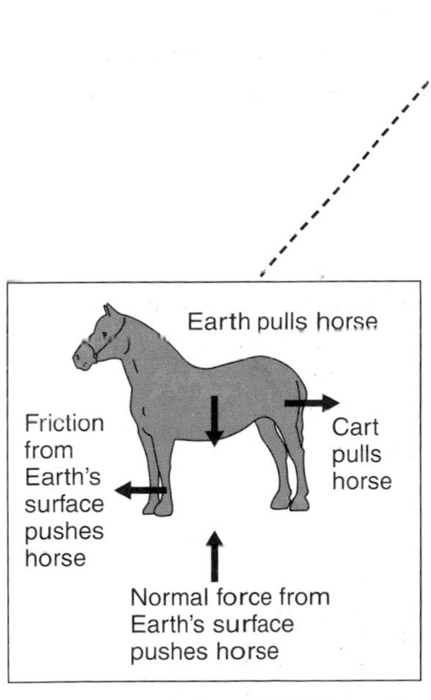

Figure 21.7 We first think about the forces on the horse. The horse pushes the Earth backwards, so the Earth pushes the horse forwards. The force of the Earth down is equal to the force of the Earth's surface upwards; and the force of the Earth to the left is equal to the force of the cart to the right.

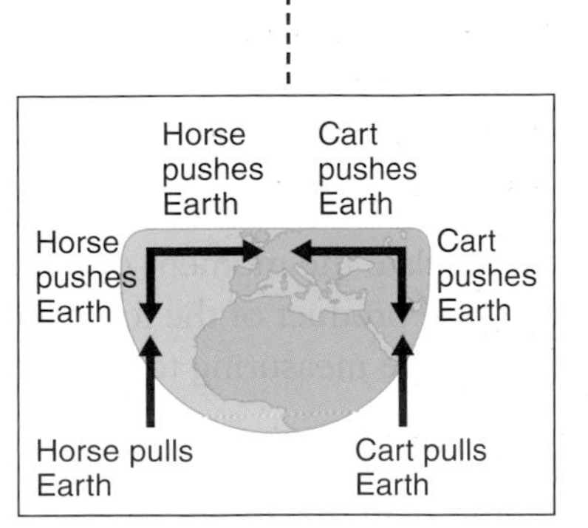

Figure 21.9 For the Earth, the sum of the forces in any direction is also zero

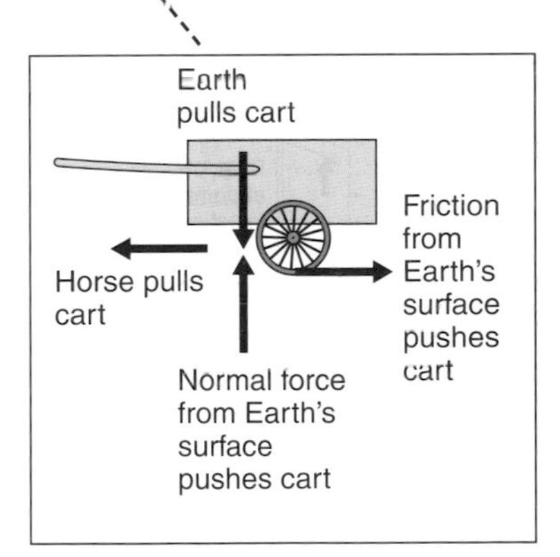

Figure 21.8 Now consider the cart. The cart is in equilibrium, so the force of the Earth down is equal to the force of the Earth's surface upwards; and the force of the horse to the left is equal to the force of the ground to the right.

26 Conservation of momentum

Figure 26.1 Gliders colliding on an air track

Collision between two bodies

If there are no external forces acting on a number of bodies, you find that momentum is conserved. This means that the total momentum after a collision is equal to the total momentum before the collision. This is the principle of **conservation of momentum**. This principle can be applied when two gliders collide on an air track (Figure 26.1).

Collisions on an air track

- Place two gliders of equal mass on an air track as shown in Figure 26.2. Give the left glider a push so that it collides and sticks to the right glider. The light gates measure the velocity of the left glider before the collision and of both gliders afterwards, by timing how long the card of known length takes to pass through the gates.
- Calculate the momentum of the left glider before the collision and the momentum of both gliders together afterwards.
- Repeat with different initial velocities and also vary the masses of the gliders involved.
- Extend your investigation to situations where the two gliders do not join together. In this case you will have to find the velocity of each glider separately after the collision. Remember that if the first glider moves to the left after the collision, then its velocity and momentum will be negative.

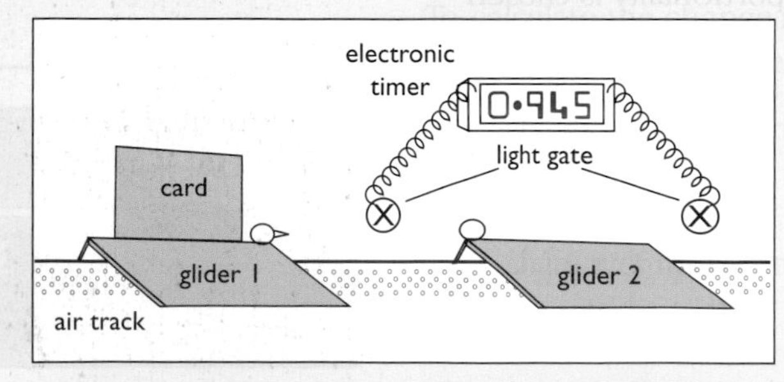

Figure 26.2 Investigating collisions on an air track

With a front-wheel-drive car (Figure 21.5), the road pushes the two front wheels forwards; it also pulls the two rear wheels backwards which results in unhelpful resistance to the car moving forwards.

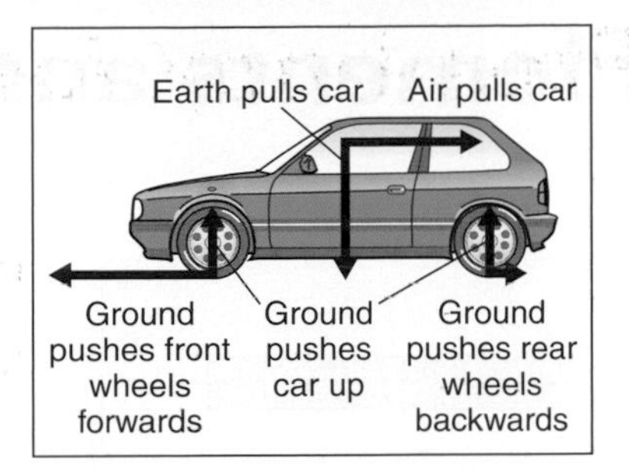

Figure 21.5 Front-wheel-drive car: In this case, only the front wheels push the ground backwards, so the ground pushes these two wheels forwards. The rear wheels are not driven; the ground pushes them backwards.

Horse pulling cart

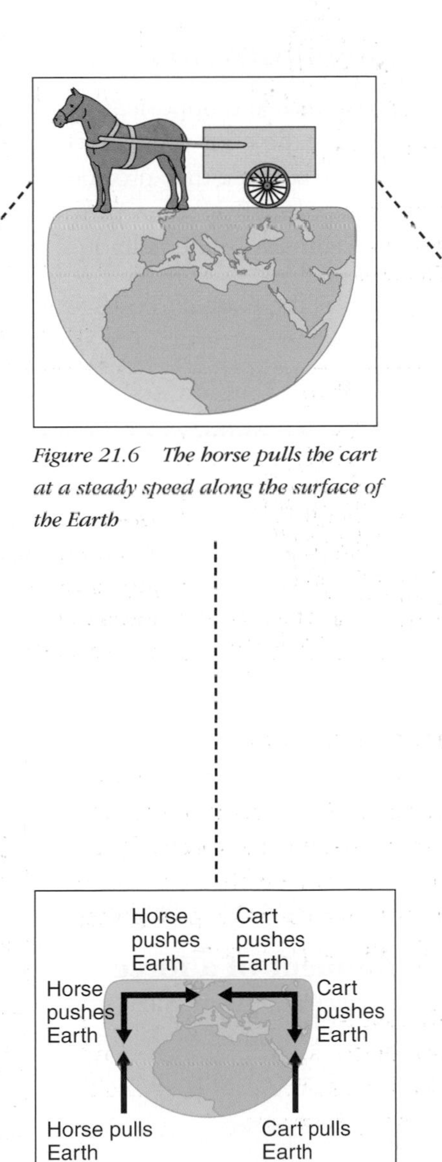

Figure 21.6 The horse pulls the cart at a steady speed along the surface of the Earth

Figure 21.9 For the Earth, the sum of the forces in any direction is also zero

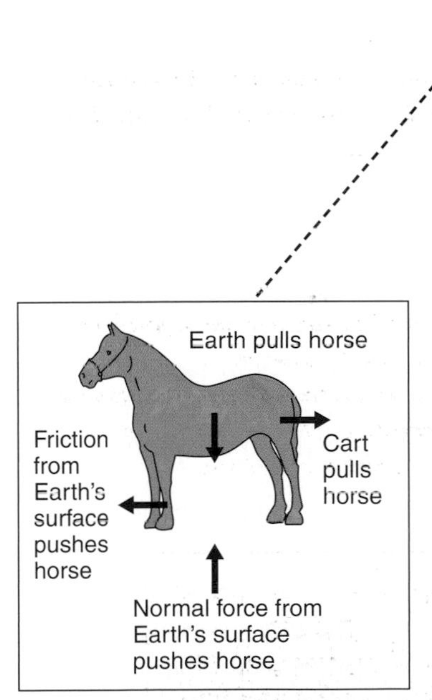

Figure 21.7 We first think about the forces on the horse. The horse pushes the Earth backwards, so the Earth pushes the horse forwards. The force of the Earth down is equal to the force of the Earth's surface upwards; and the force of the Earth to the left is equal to the force of the cart to the right.

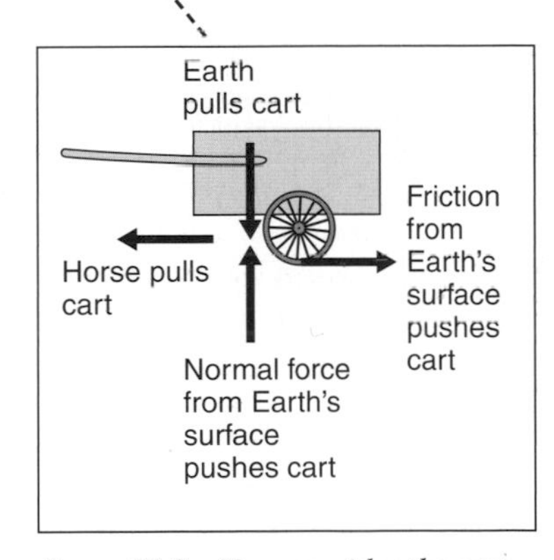

Figure 21.8 Now consider the cart. The cart is in equilibrium, so the force of the Earth down is equal to the force of the Earth's surface upwards; and the force of the horse to the left is equal to the force of the ground to the right.

22 Moments and couples

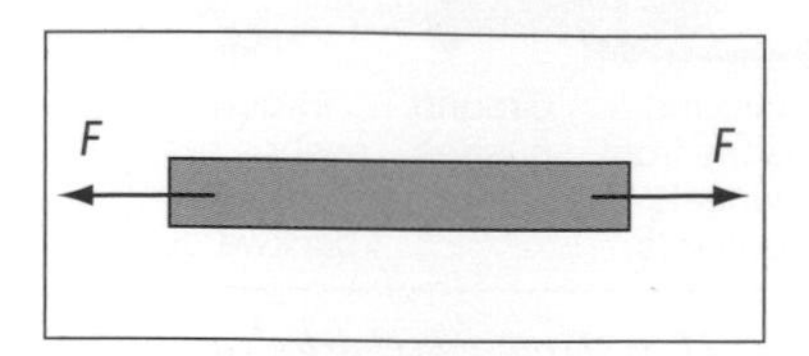

Figure 22.1 This rod is under tension

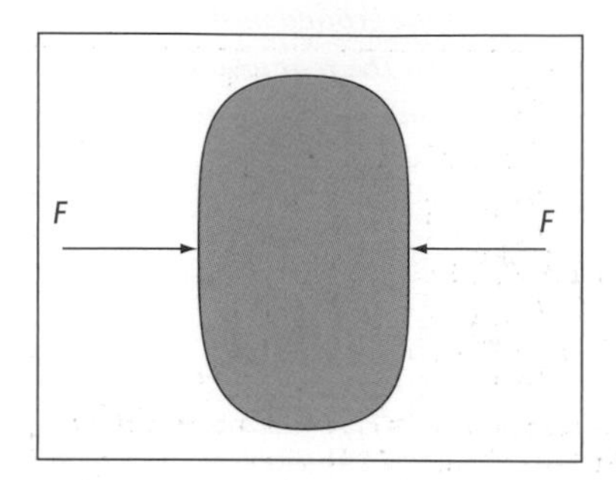

Figure 22.2 This ball is under compression

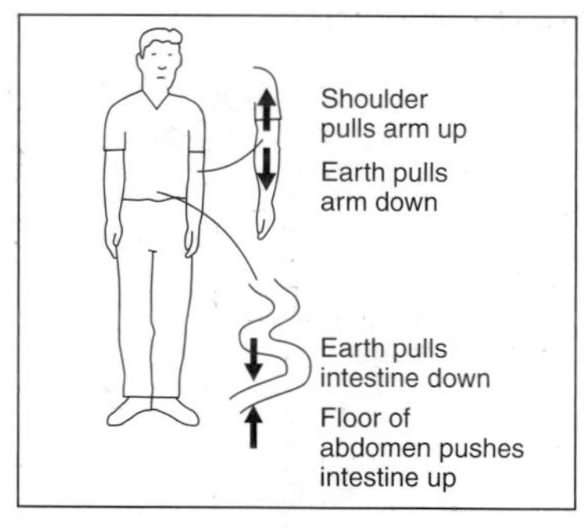

Figure 22.3 Gravity and supporting forces put some parts of your body into tension and some into compression

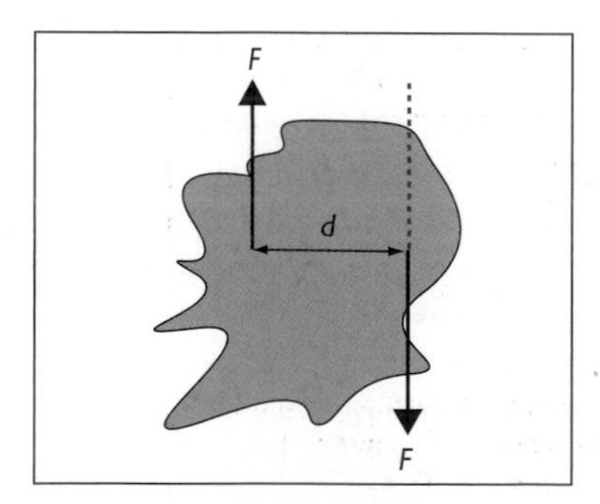

Figure 22.4 This pair of forces is a couple, which causes rotation

Two forces

Very often, a body has a pair of equal and opposite forces acting on it. If the weight of the rod in Figure 22.1 is so little that you can ignore it, the rod will be in equilibrium if the leftward force is equal to the rightward force. Even though the rod is in equilibrium, it is still in a state of **tension** caused by forces that stretch it. The ball in Figure 22.2 is also in equilibrium. But it is in **compression** caused by the pair of forces that squash it. Parts of your body are in compression and other parts are in tension (Figure 22.3).

Turning moments

If a pair of equal and opposite forces, which are not in line, act on a body, they tend to cause the body to rotate (Figure 22.4). The forces cause no linear acceleration, because they provide equal and opposite linear effects. But they have a rotational effect. They provide a **couple**. The body on which they act will be in *translational* equilibrium, but not in *rotational* equilibrium.

The ability of the couple to cause rotation depends on the size of the forces and how far they are apart. The size of this effect is called the turning moment or just the moment.

You can find the **moment of a couple** by multiplying one of the forces F by the perpendicular distance d between them. In Figure 22.4, the moment = Fd. Moments of different sizes are shown in Figure 22.5.

Moments cause things to rotate about a line called the **axis**. For instance, if you turn a steering wheel, your hands are producing a moment about the steering column, the rod that joins the wheel to the rest of the steering mechanism.

Turning moment is calculated by multiplying a force by a distance which is perpendicular to the line of action of the force. Moment is measured in newton metres (N m). The physical quantity of work is also calculated by multiplying a force by a distance, but in this case the force is multiplied by a distance in the direction of the force and also has units of newton metres (see Chapter 28). The unit of work is always called the joule and the unit of moment is always referred to as a newton metre.

The moment of a force

You can calculate the moment of a single force about any axis. The **moment of a force** is the product of the force and the perpendicular distance from the axis that you are measuring to (Figure 22.6).

Torque

Often, more than two forces cause a turning moment; for instance, more than four different forces, at different times, cause the turning moment on the crankshaft of a car. A **torque** is the turning moment caused by a system of two or more forces that tend to cause rotation.

Neither a couple nor a torque has any resultant force in any direction, so neither causes any linear acceleration. They have no translational effect.

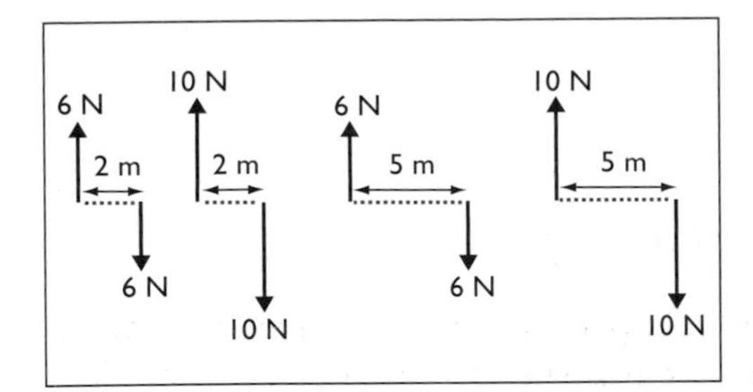

Figure 22.5 The largest moments occur with large forces and a large perpendicular distance

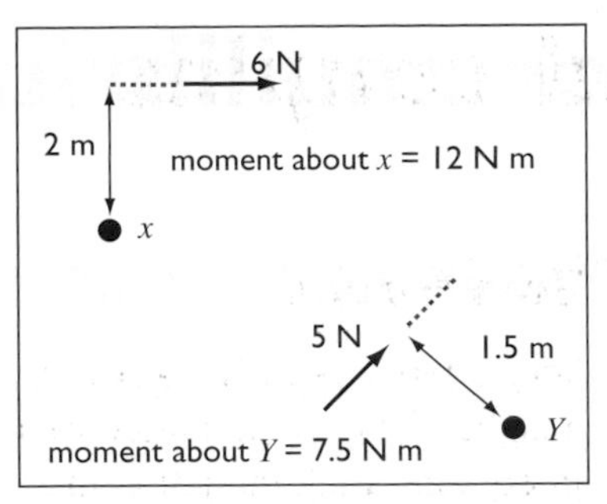

Figure 22.6 The moment of a force about a point

Principle of moments

If a body is in equilibrium, you know that the sum of the forces in any direction must be zero. There is another condition for equilibrium. If a body is in equilibrium, the sum of the moments turning the body one way must equal the sum of the moments turning it in the opposite direction. The sum of the moments about any axis must be zero. This is the **principle of moments**. The moments for a father and child balanced on a see-saw are shown in Figure 22.7.

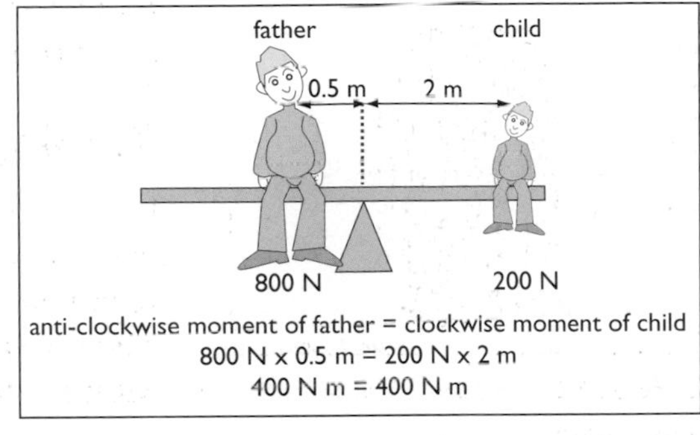

Figure 22.7 The father's anticlockwise moment equals the child's clockwise moment

Forces on bridge supports

- A lorry on a bridge can be modelled using apparatus set up as in Figure 22.8. Use the model bridge to investigate how the forces on the two supports change as a lorry moves over a bridge.
- Set up the apparatus as shown. Measure the forces on the two supports as the lorry moves over the bridge.
- For each position of the lorry, calculate the moment of the forces about each support (Figure 22.9), the sum of the upward forces and the sum of the downward forces.

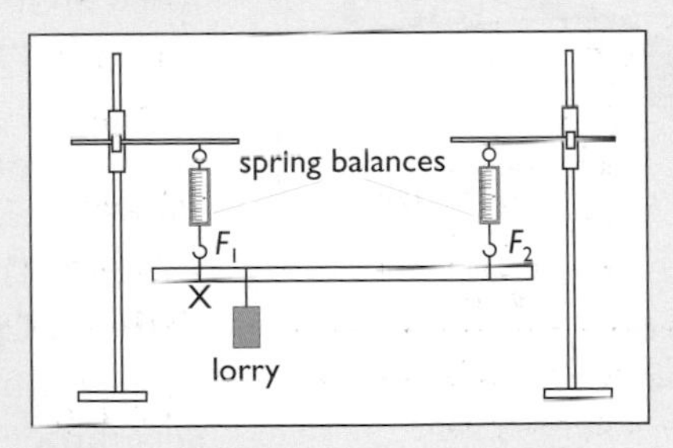

Figure 22.8 A model for a lorry on a bridge

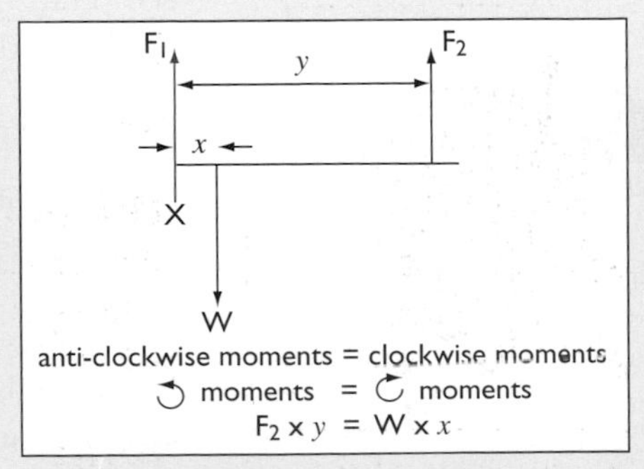

Figure 22.9 The moments about the support X

23 Conditions for equilibrium

Finding the centre of gravity of a body

- Hang the body from a suitable point. Draw a vertical line through the point of support when the body is balanced (Figure 23.1).
- Repeat with another balancing point. Where the two lines intersect is the centre of gravity (Figure 23.2).
- Repeat again with a third line to check.

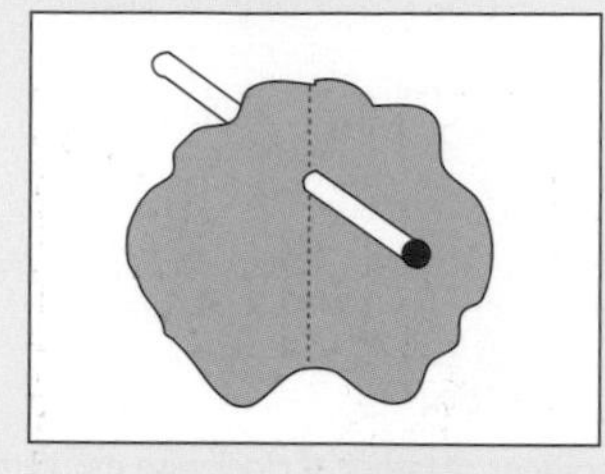

Figure 23.1 A body hanging from a point

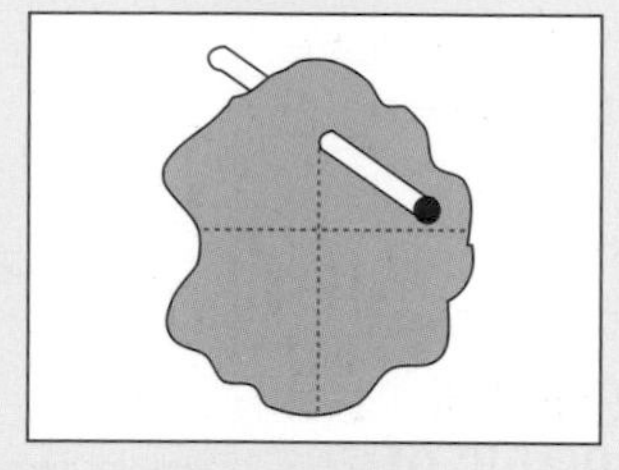

Figure 23.2 The body hanging from a new point

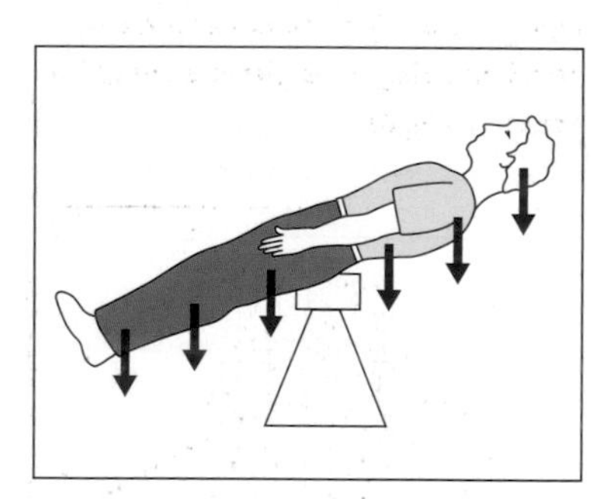

Figure 23.3 Gravity acts on all parts of a body

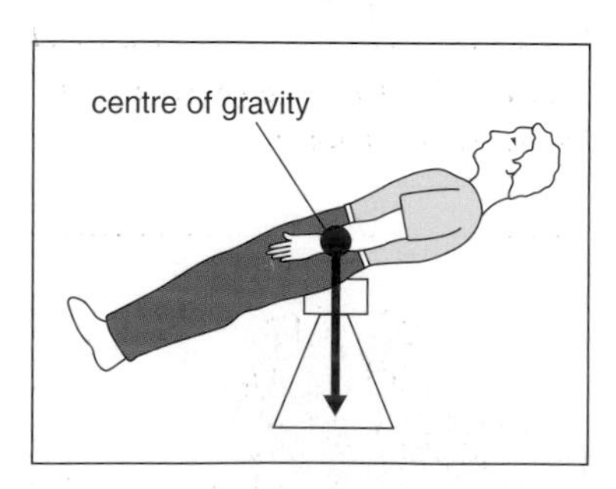

Figure 23.4 You can consider that all the weight acts through the centre of gravity

Centre of gravity

When a body is in a gravitational field, each part of the body has a gravitational force on it (Figure 23.3). If you support a body, each of these gravitational forces has a moment about the point of support.

With the support in the right place, the body is balanced. The sum of all the moments of the different parts of the body is then zero about the point of support. This occurs when the support is vertically in line with the body's **centre of gravity**, the point at which all the weight appears to act. For balancing and for linear motion you can consider that a body's whole weight acts through the centre of gravity (Figure 23.4).

For spherically symmetrical bodies, the centre of gravity is always in the centre. For other rigid bodies, you can consider the centre of gravity to be fixed, even though it depends a little on the orientation of the body.

If a body changes its shape, the centre of gravity will change position. If you bend forwards at your waist, your centre of gravity will be somewhere outside your body.

Sometimes you will encounter the term **centre of mass**. This is the place at which all the mass appears to be. For gravitational purposes, the centre of mass is in the same position as the centre of gravity, and you can treat the two terms as if they were the same.

Conditions for equilibrium

Statics is the study of forces in systems that are stationary, whereas *dynamics* is the study of forces in systems that are moving. In A-level physics, the only statics situations that you need to know about are those in which the forces are *coplanar*, which means that all the forces are confined to a single plane.

When a body is in equilibrium, the sum of the forces in any direction is zero and the sum of the moments about any axis is zero. These conditions mean that you can usually write down three equations:

- equation (1) states that forces in any one direction are zero
- equation (2) states that forces in any other direction are zero
- equation (3) states that the moments about any point are zero.

You can choose the directions and the point to make things easy: the two directions are usually horizontal and vertical. It is often helpful to choose a point through which at least two forces pass to simplify the moments equation.

For example, Figure 23.5 shows a free-body force diagram for a ladder resting against a wall. In this case your three equations could be:

- equation (1): the resultant vertical force must be zero
- equation (2): the resultant horizontal force must be zero
- equation (3): the anticlockwise moment of R_1 about the point of contact of the ladder with the floor must equal the clockwise moment of the weight about the same point.

So the equations would be:

$$R_2 - 200\ \text{N} = 0 \qquad (1)$$

$$R_1 - F = 0 \qquad (2)$$

$$R_1 \times 3\ \text{m} = 200\ \text{N} \times 1.2\ \text{m} \qquad (3)$$

From equation (1), you can calculate that $R_2 = 200$ N.

From equation (3), you can calculate that $R_1 = 80$ N.

So from equation (2), you can calculate that $F = 80$ N.

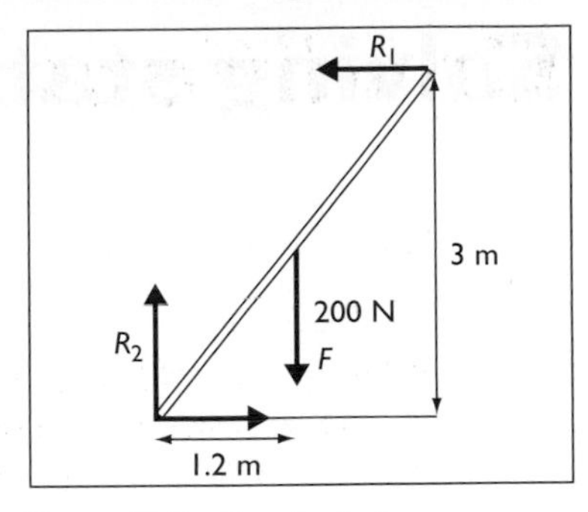

Figure 23.5 Free-body force diagram for a ladder resting against a wall

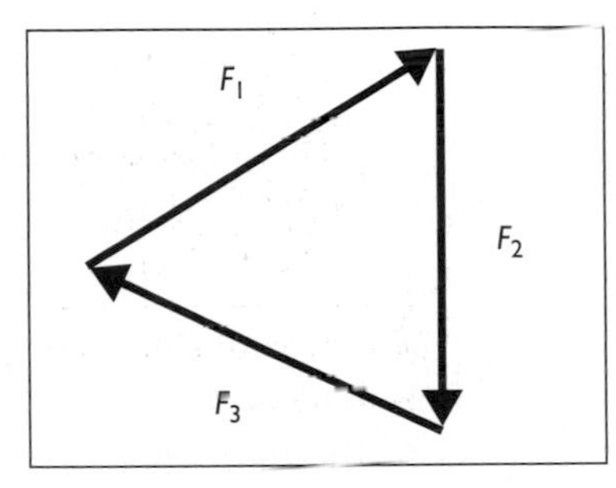

Figure 23.6 If three forces have no resultant, they must form a complete vector triangle

Special rules for situations involving three forces

When a body is in equilibrium under the action of three forces that are not parallel, two useful rules apply:

(i) When you draw the forces head-to-tail one after another, they must form a closed triangle (Figure 23.6). Drawing them head-to-tail is the same as adding them with a vector diagram. If a number of forces have no resultant, when you draw them head-to-tail, the end of the last one must be at the start of the first.

(ii) All three forces must pass through a single point (Figure 23.7). If they don't pass through the same point, the body is *not* in equilibrium (Figure 23.8).

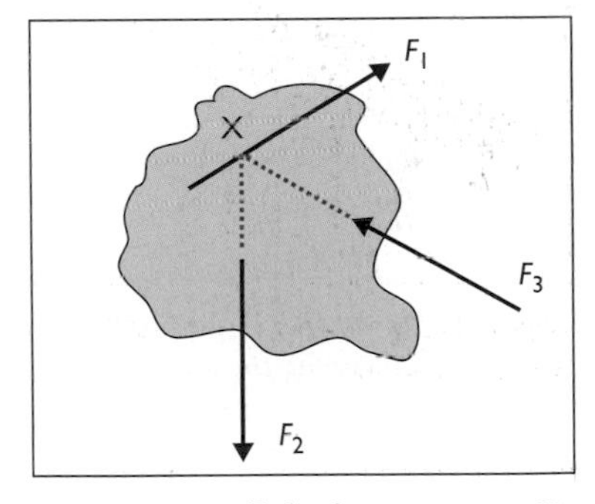

Figure 23.7 All the forces cross at X, so there is no moment

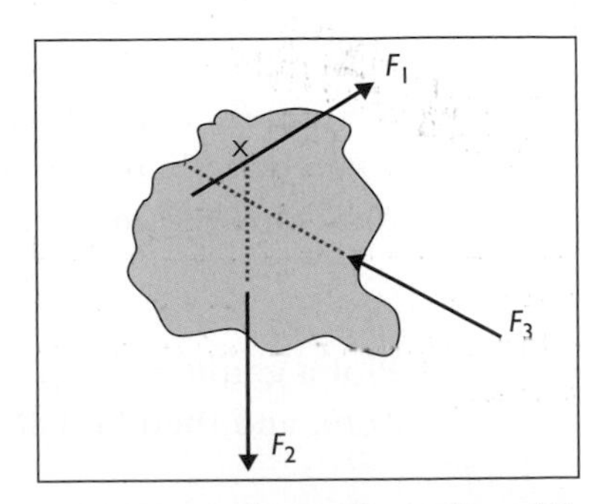

Figure 23.8 The two forces F_1 and F_2 cross at X, but F_3 does not. So F_3 has a moment about X and the body is not in equilibrium

24 Solving statics problems

Answering statics problems

There are many situations in which you need to find two or three unknown forces on a body in equilibrium, on which one known force is acting. Here are some hints.

1 Look for the simplest of the unknown forces.
2 Take moments about a point through which the other one or two unknown forces pass.
3a EITHER: resolve horizontally and vertically to find the other unknown(s).
3b OR: take moments about another point to calculate the other unknown(s).

Finding the mass of a retort stand

- Balance a retort stand on its own to find its centre of gravity (Figure 24.1).
- Then, with the help of a set of masses, balance it about a different point (Figure 24.2).
- Calculate the clockwise moments of the set of masses. This is equal to the anticlockwise moment of the mass of the retort stand.
- From this, calculate the mass of the retort stand.

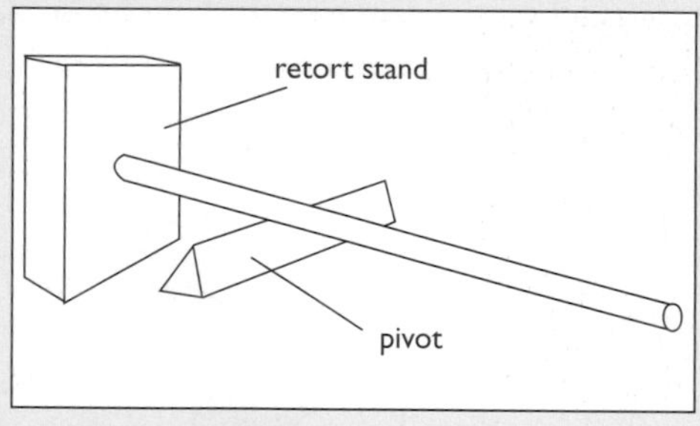

Figure 24.1 Find the centre of gravity by balancing the retort stand alone

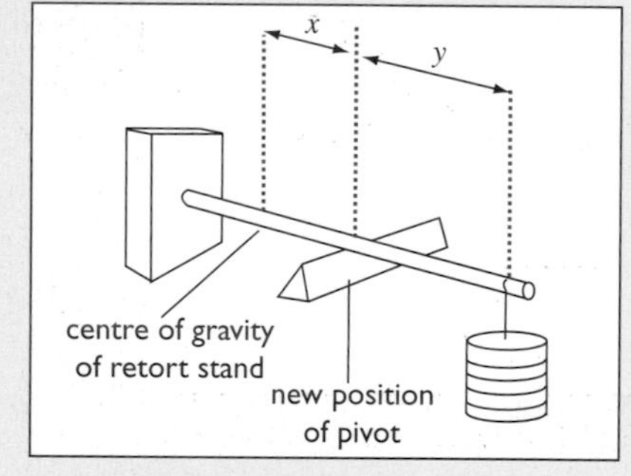

Figure 24.2 Then balance the retort stand with a known mass on the end

Measuring an unknown mass

- Set up the apparatus as in Figure 24.3.
- For a range of positions of the unknown mass M, find where you should put the known mass m to balance it.
- Anticlockwise moments = clockwise moments

$$Mgx = mgy$$
$$y = (M/m)x$$

- Plot a graph of y against x. Measure the slope, which is M/m, and then find M.

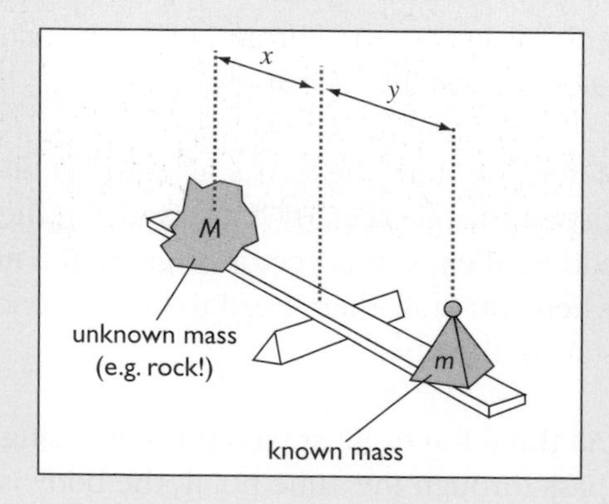

Figure 24.3 You can use the principle of moments to measure an unknown mass

WORKED EXAMPLE

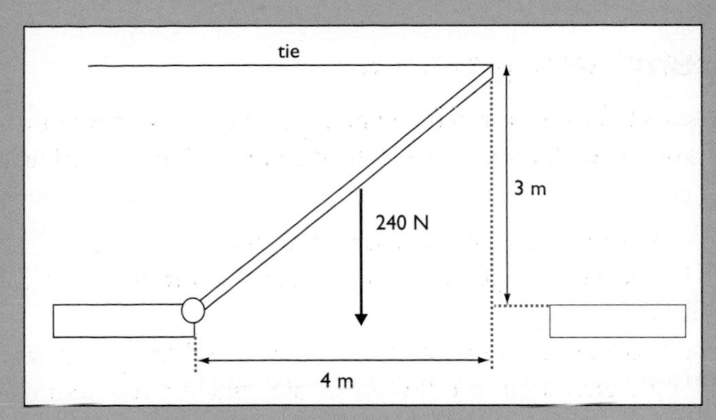

Figure 24.4 Trapdoor

Figure 24.4 shows a trapdoor of weight 240 N held up by a horizontal tie. Calculate the tension in the tie and the horizontal and vertical components of the force at the hinge.

First draw a free-body force diagram for the trapdoor, as in Figure 24.5.

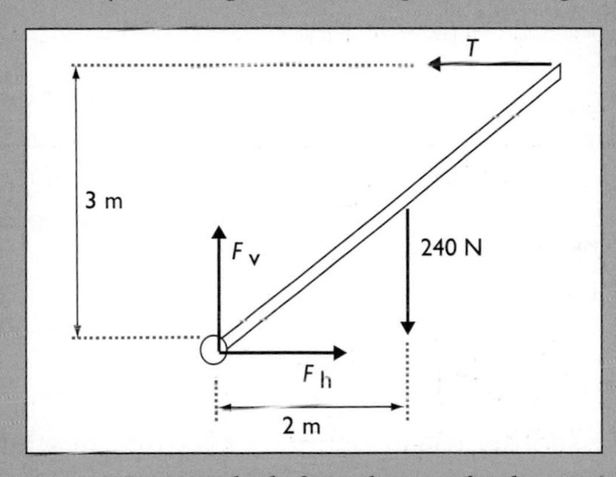

Figure 24.5 Free-body force diagram for the trapdoor

The force at the hinge is represented by its vertical component F_v and horizontal component F_h. The only vertical forces are F_v and the weight. So F_v must be 240 N.

The simplest way to find T is to take moments about the hinge. Both F_v and F_h act through the hinge, and therefore have no moment about it. So

$$\begin{aligned} \text{anticlockwise moments} &= \text{clockwise moments} \\ T \times 3\text{ m} &= 240\text{ N} \times 2\text{ m} \\ T &= \frac{480\text{ N}}{3} \\ &= 160\text{ N} \end{aligned}$$

The only horizontal forces are F_h and the tension T, so F_h must be 160 N.

25 Momentum

Momentum

Momentum is a physical quantity that allows you to do calculations about what happens when moving bodies collide. You can use momentum to help you to analyse a car crash, the impact of a racket on a ball or, a collision between an air-rifle pellet and an air track glider. (The momentum in this course is strictly called *linear* momentum, to distinguish it from the *angular* momentum of a spinning body.)

The **momentum** of a body is given by the word equation:

momentum = mass × velocity

or in symbols:

$$p = mv$$

You can see from the equations that the unit of momentum is the kg m s^{-1}.

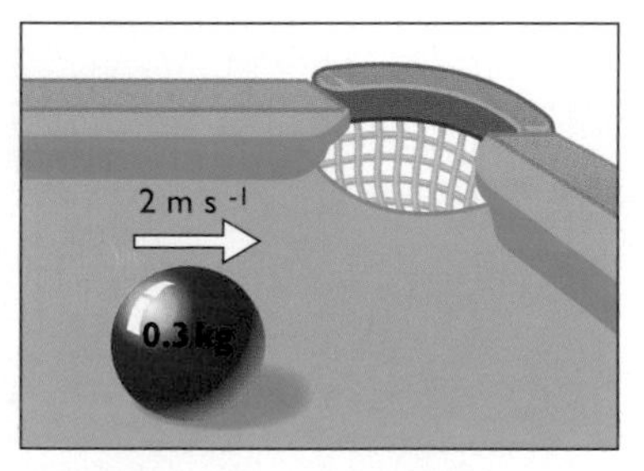

Figure 25.1 Momentum of snooker ball moving to right. Momentum = 0.3 kg × 2 m s^{-1} = 0.6 kg m s^{-1} to right

Momentum is a vector quantity; its direction is the same as that of the velocity of the body. The momentum of the snooker ball in Figure 25.1 is 0.6 kg m s^{-1} to the right. Figure 25.2 shows the ball after it has collided with the cushion. Its momentum is now 0.3 kg m s^{-1} to the left, which you could express as –0.3 kg m s^{-1} to the right.

Change of momentum

The momentum of the snooker ball in Figure 25.1 is 0.6 kg m s^{-1}. Its momentum in Figure 25.2, after colliding with the cushion, is –0.3 kg m s^{-1}. To calculate the change in momentum, as with any change, subtract the initial value from its final value. So

change in momentum = final momentum – initial momentum

In this case

$$\text{change of momentum} = (-0.3\ \text{kg m s}^{-1}) - (0.6\ \text{kg m s}^{-1})$$
$$= -0.9\ \text{kg m s}^{-1}$$

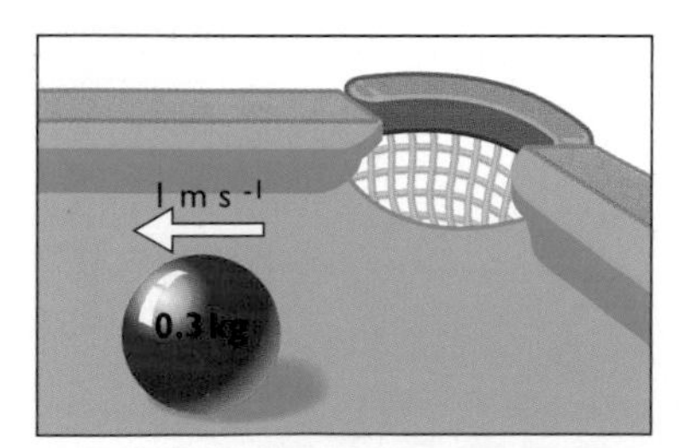

Figure 25.2 Momentum of snooker ball moving to left. Momentum = 0.3 kg × 1 m s^{-1} = 0.3 kg m s^{-1} to left

The change of momentum is negative, when rightwards is positive; this means that the change of momentum is to the left. The change of momentum of the snooker ball is caused by the force of the cushion on the ball during the collision.

Momentum and Newton's second law

In Chapter 13 you learned a statement of Newton's second law connecting force and acceleration. Newton himself expressed his second law in terms of momentum stating that:

> The rate of change of momentum of a body is directly proportional to the resultant force acting on it and takes place in the same direction as the resultant force.

This version of **Newton's second law** defines **force** to be the quantity which causes a rate of change of momentum. The larger the force, the larger the rate of change of momentum, or, if you prefer, the larger the change of momentum per second (Figure 25.3). The law also makes an important statement about direction. The rate of change of momentum is in the same direction as the force; that is, if the force is to the left, the rate of change of momentum is towards the left.

Figure 25.3 A large force changes the momentum of a golf ball in a short time

From this version of Newton's second law you can state:

$$\text{force} \propto \text{rate of change of momentum}$$

Since *rate of* means *divided by time*, and change = final – initial,

$$\text{force} \propto \frac{(\text{final momentum} - \text{initial momentum})}{\text{time}}$$

If the mass m is constant you can express this in symbols:

$$F \propto \frac{(mv - mu)}{t}$$

where u and v are the initial and final velocities, and t is the time between these velocities.

In the International System of Units, the constant of proportionality is chosen to be 1. Therefore

$$F = \frac{(mv - mu)}{t}$$

$$F = \frac{m(v - u)}{t}$$

Since $(v - u)/t$ is the acceleration a, we get

$$F = ma$$

26 Conservation of momentum

Figure 26.1 Gliders colliding on an air track

Collision between two bodies

If there are no external forces acting on a number of bodies, you find that momentum is conserved. This means that the total momentum after a collision is equal to the total momentum before the collision. This is the principle of **conservation of momentum**. This principle can be applied when two gliders collide on an air track (Figure 26.1).

Collisions on an air track

- Place two gliders of equal mass on an air track as shown in Figure 26.2. Give the left glider a push so that it collides and sticks to the right glider. The light gates measure the velocity of the left glider before the collision and of both gliders afterwards, by timing how long the card of known length takes to pass through the gates.
- Calculate the momentum of the left glider before the collision and the momentum of both gliders together afterwards.
- Repeat with different initial velocities and also vary the masses of the gliders involved.
- Extend your investigation to situations where the two gliders do not join together. In this case you will have to find the velocity of each glider separately after the collision. Remember that if the first glider moves to the left after the collision, then its velocity and momentum will be negative.

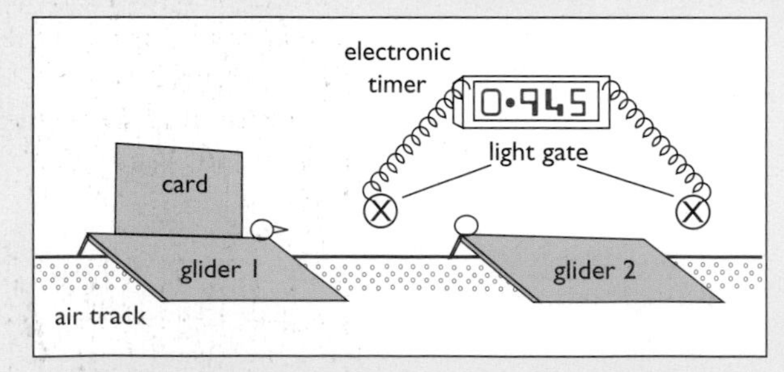

Figure 26.2 Investigating collisions on an air track

Explosions

When a crossbow shoots an arrow (Figure 26.3), the bow pushes the arrow forwards and the arrow pushes the bow backwards. Momentum is conserved. So the momentum gained in a forward direction by the arrow is equal to the backward momentum that the arrow gives to the bow and the archer.

Figure 26.3 A bow firing an arrow

When you fire an air-rifle (Figure 26.4), a sudden explosion pushes a pellet (and some gas) forwards at the same time as pushing the air-rifle backwards. The forward momentum that the explosion gives to pellet and gas is equal to the backward momentum that it gives to the air-rifle.
Similarly, when bombs and shells explode, the increase in momentum in any one direction is matched by an increase in momentum in the opposite direction, so the total momentum is unchanged.

A rocket engine fires hot gases from its exhaust. As the engine pushes the gases backwards, the gases push the rocket forwards (Figure 26.5). Again, the increase in the forward momentum of the rocket is equal to the increase in the backward momentum of the exhaust gases.

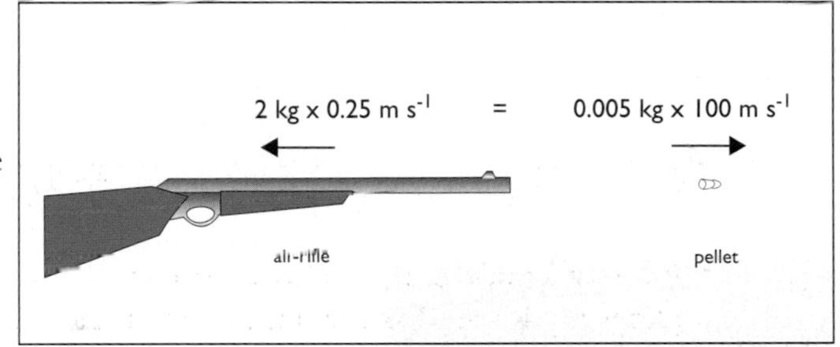

Figure 26.4 If you hold the air-rifle so that it can move backwards freely, you can assume that the momentum of the pellet is equal and opposite to that of the air-rifle

Figure 26.5 European rocket launcher taking off

Table 26.1 *Conservation*

Conservation often means 'keeping from loss or keeping from damage'. For instance, people want to conserve rain forests so that we don't lose them, or conserve the environment so that we don't lose any species of life that live here.

Physicists use the word *conservation* to mean 'keeping from change'. The law of conservation of momentum means that for a closed system there is *no* change in momentum, so that the momentum before a collision is equal to the momentum after the collision.

27 Impulse

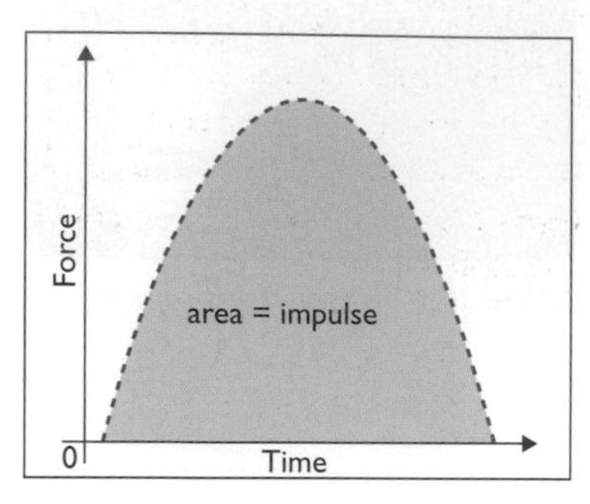

Figure 27.1 The area under a force–time graph gives the impulse

Impulse

The same change in momentum can be caused by a small force acting on a body for a long time, or a larger force acting on the body for a shorter time. Since

$$\text{force} = \frac{\text{change of momentum}}{\text{time}}$$

$$\text{force} \times \text{time} = \text{change of momentum}$$

Or in symbols,

$$Ft = mv - mu$$

The product force × time is the **impulse**; it is equal to the change of momentum. Impulse depends on both the size of the force and the time for which it acts.

Figure 27.2 A small gravitational force from the Sun, acting over a long time, gives the comet enormous momentum

It is easy to calculate the impulse of a steady force: simply multiply the force by the time for which it acts. If the force is changing steadily, multiply the average force by the time. If you have a graph of force against time, the impulse is equal to the area under the graph (Figure 27.1).

Since impulse is equal to change of momentum, the units of both must be the same: the kg m s^{-1}. You can see that the units of impulse are also the units of force × time, which is the N s. These are two different ways of saying the same thing, since

$$\text{N s} = \text{kg m s}^{-2} \times \text{s} = \text{kg m s}^{-1}$$

You can use this unit for both change of momentum and impulse. However, it is common practice to state impulse in N s and momentum in kg m s^{-1}.

The Sun pulls the comet in Figure 27.2 with a small (gravitational) force acting for a very long time.

Measuring the force of a kick

- Attach foil to both a football and the shoe of a kicker and use an electronic timer to measure the time they are in contact during a kick (Figure 27.3).
- Kick the ball horizontally. Measure the vertical distance fallen, the horizontal distance travelled and the duration of the kick.
- Measure the mass of the football and calculate the horizontal velocity of the ball as described in Chapter 11 (Practical: Measuring the speed of a snooker ball). Multiply mass by velocity to calculate the momentum given, and therefore the impulse applied to the ball.
- Divide the impulse by the time of contact to calculate the average force applied.

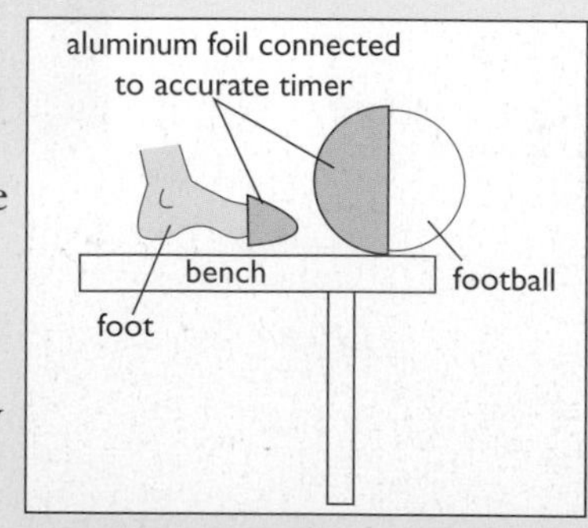

Figure 27.3 Measuring the contact time between ball and foot

Impulses in collisions

When two bodies collide, they exert equal and opposite forces on each other (Figure 27.4) and these Newton III pair of forces act for the same length of time.

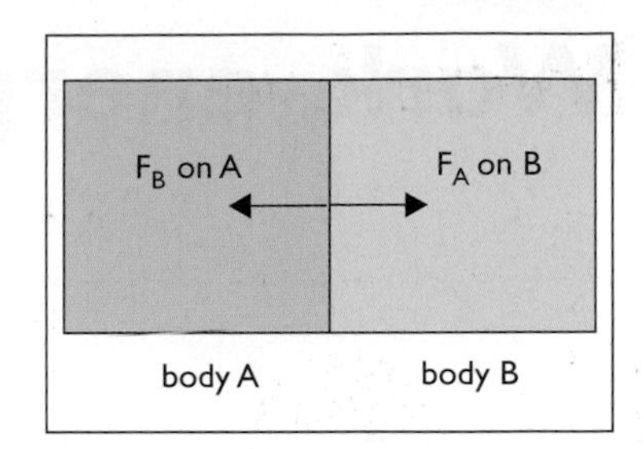

Figure 27.4 The forces between two colliding bodies

Since the times are equal and the forces are equal and opposite, the impulses are equal and opposite. Body A exerts on body B an equal and opposite impulse to that which body B exerts on body A. Therefore the change in the momentum of body B is equal and opposite to the change in the momentum of body A.

So the overall change in the total momentum of both bodies during the collision is zero; that is, momentum is conserved.

WORKED EXAMPLE

A model railway truck of mass 2 kg moving at 5 m s^{-1} collides with another truck of mass 6 kg moving at 1 m s^{-1} in the same direction. If the two trucks join together, find their speed after the collision and the impulse exerted by one truck on the other.

Draw a diagram of the situation, as in Figure 27.5.
Then use conservation of momentum,

$$\text{initial momentum} = \text{final momentum}$$

and using rightwards as positive, this is

$$(2\text{ kg} \times 5\text{ m s}^{-1}) + (6\text{ kg} \times 1\text{ m s}^{-1}) = 8\text{ kg} \times v$$
$$10\text{ kg m s}^{-1} + 6\text{ kg m s}^{-1} = 8\text{ kg} \times v$$
$$16\text{ kg m s}^{-1} = 8\text{ kg} \times v$$
$$v = 2\text{ m s}^{-1}$$

Figure 27.5 Before and after a collision

So both trucks end up moving at 2 m s^{-1} to the right.

$$\text{impulse on 2 kg mass} = mv - mu = (2\text{ kg} \times 2\text{ m s}^{-1}) - (2\text{ kg} \times 5\text{ m s}^{-1})$$
$$= 4\text{ kg m s}^{-1} - 10\text{ kg m s}^{-1} = -6\text{ kg m s}^{-1}$$
$$\text{impulse on 6 kg mass} = (6\text{ kg} \times 2\text{ m s}^{-1}) - (6\text{ kg} \times 1\text{ m s}^{-1})$$
$$= 12\text{ kg m s}^{-1} - 6\text{ kg m s}^{-1} = +6\text{ kg m s}^{-1}$$

If the same collision happens, but this time the 6 kg truck bounces away from the collision at 3 m s^{-1} in the same direction, find the final speed of the 2 kg truck.
Now look at Figure 27.6.
Again

$$\text{initial momentum} = \text{final momentum}$$

and using rightwards as positive,

$$(2\text{ kg} \times 5\text{ m s}^{-1}) + (6\text{ kg} \times 1\text{ m s}^{-1}) = (2\text{ kg} \times v) + (6\text{ kg} \times 3\text{ m s}^{-1})$$
$$16\text{ kg m s}^{-1} = 2\text{ kg} \times v + 18\text{ kg m s}^{-1}$$
$$16\text{ kg m s}^{-1} - 18\text{ kg m s}^{-1} = 2\text{ kg} \times v$$
$$-2\text{ kg m s}^{-1} = 2\text{ kg} \times v$$
$$v = -1\text{ m s}^{-1}$$

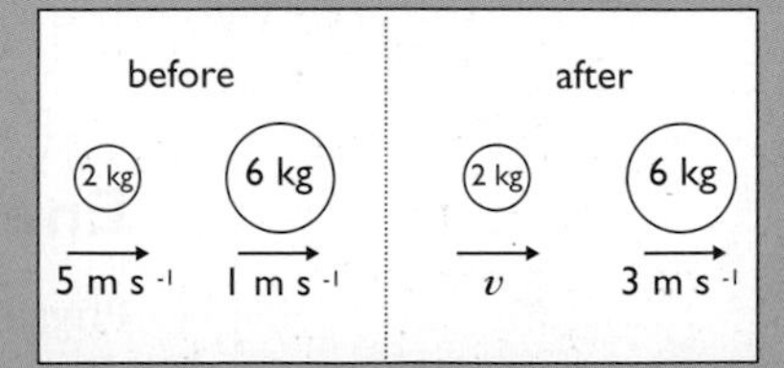

Figure 27.6 Before and after another collision

This means that the 2 kg truck moves at 1 m s^{-1} to the left.

28 Work, energy and power

Work

Many tasks involving force or motion are tiring. For instance, if you were supporting a heavy load you would get tired and need an input from your food. But you could be replaced by a shelf which would support the load without needing to be fed (Figure 28.1).

Figure 28.1 Two methods of supporting the same load

If a moving object were on a smooth surface, or in outer space, it would continue to move at a constant speed. It doesn't need an input to keep it going.

But there are many tasks that by their very nature do need an input. For instance, if you want to raise a load, or grind a mineral, or drill a hole, you need a person or animal that needs feeding, or something like an electric motor or a petrol-powered engine that needs electricity or fuel. Processes that need an input like this are what physicists call **work**. Work is done when a body exerts a force and moves a distance in the direction of the force:

work = force × displacement in the direction of the force

$$\Delta W = F\,\Delta x$$

A small bunch of keys may have a weight of about 1 N. If you raise them through a height of 1 m

$$\Delta W = F\,\Delta x = 1\ \text{N} \times 1\ \text{m} = 1\ \text{N m}$$

The unit of work is the newton metre (N m), called a joule (J).

Another example is shown in Figure 28.2.

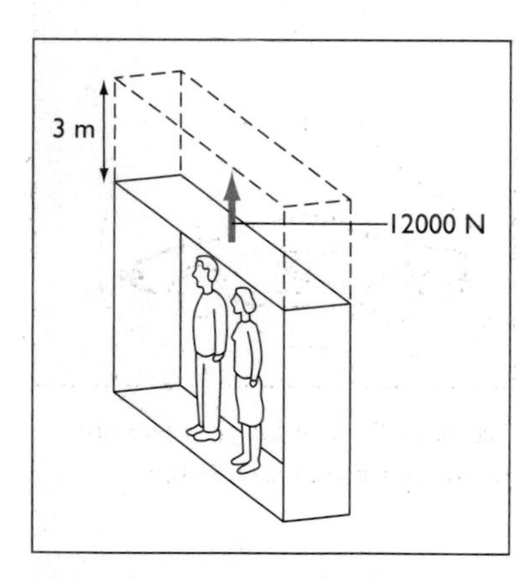

Figure 28.2 Raising this lift 3 m needs 36000 J

If the force is changing steadily, you can calculate the work done by multiplying the average force by the displacement in the direction of the force. If you have a graph of force against displacement, the work done is the area under the graph (Figure 28.3).

Figure 28.4 shows how to calculate the work done when the face is at an angle θ to the displacement:

$$W = (F \cos \theta)\,\Delta x$$

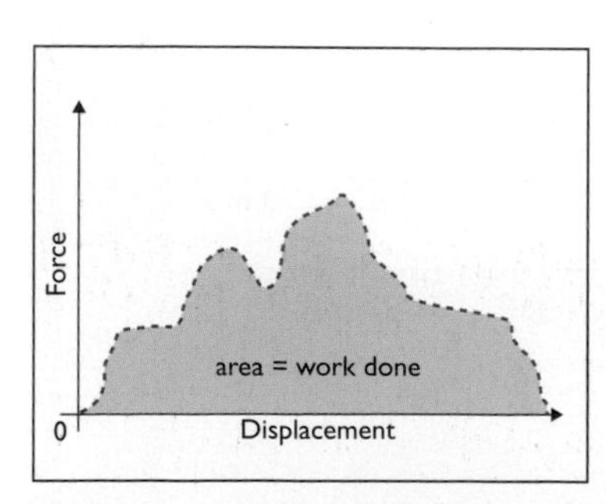

Figure 28.3 The area under a force–displacement graph gives the work done

Energy

Physicists use the word **system** to describe a body, or a group of bodies, that they are thinking about. **Energy** is a mathematical quantity which changes when work is done on, or by, a system (Figure 28.5). When a system does work, its energy decreases. When you do work on a system, its energy increases.

If you do work squashing a spring, your energy decreases, and the energy of the spring increases. If a battery-powered winch raises a load, then the energy of the battery decreases and the energy of the load increases.

If any system does an amount of work ΔW, then its energy decreases by an amount ΔW, and the energy of the system that it is working on increases by an amount ΔW. The amount of energy this transfers from one system to the other is equal to the amount of work done.

energy transferred = work done

So if you do 250 J of work squashing a spring, your energy decreases by 250 J and the energy of the spring increases by 250 J.

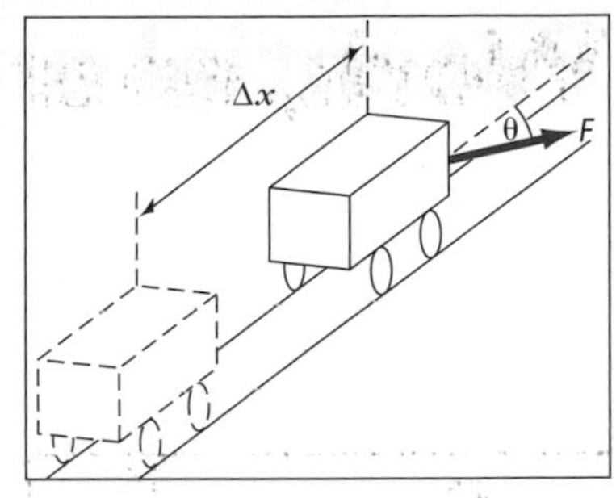

Figure 28.4 If the force is at an angle θ to the displacement, multiply the component of the force in the direction of the displacement (F cos θ) by Δx

Power

The rate of doing work is called **power**. This is the same as the rate of energy transfer. You know that rate of means divided by time. So you can see that

$$\text{power} = \frac{\text{energy transfer}}{\text{time}} = \frac{\text{work}}{\text{time}}$$

In symbols,

$$\text{power} = \frac{\Delta W}{\Delta t}$$

where the use of the Δ symbols means that the times are short.

The unit of power is the joule per second or watt (W).

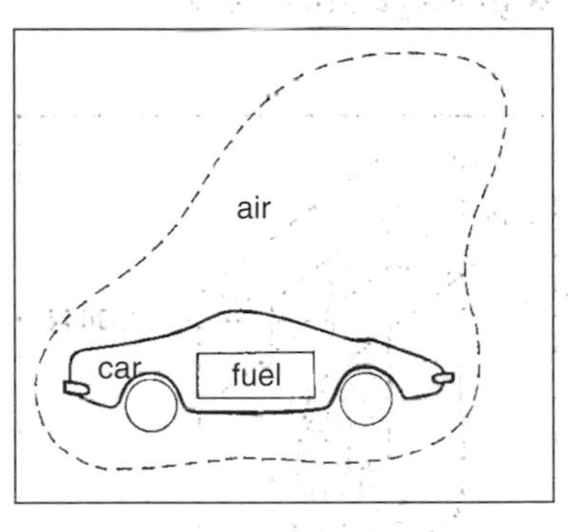

Figure 28.5 The car, fuel and air are a system that has energy

In the days when horsepower was important, it was reckoned that a horse could raise a weight of 500 N when moving 3 m in 2 s. The work done in 2 s is

$$F\,\Delta x = 500\text{ N} \times 3\text{ m} = 1500\text{ J}$$

so that

$$\text{power} = \frac{\text{work}}{\text{time}} = \frac{1500\text{ J}}{2\text{ s}} = 750\text{ W}$$

Here is another equation for power.
Since power = work/time and work = force × displacement, then

$$\text{power} = \frac{\text{work}}{\text{time}} = \frac{\text{force} \times \text{displacement}}{\text{time}} = \text{force} \times \frac{\text{displacement}}{\text{time}}$$

$$\text{power} = \text{force} \times \text{velocity}$$

The horse above pulls a load of 500 N at a speed of 1.5 m s^{-1}. Its power is therefore 500 N × 1.5 m s^{-1} = 750 W.

Measuring your power

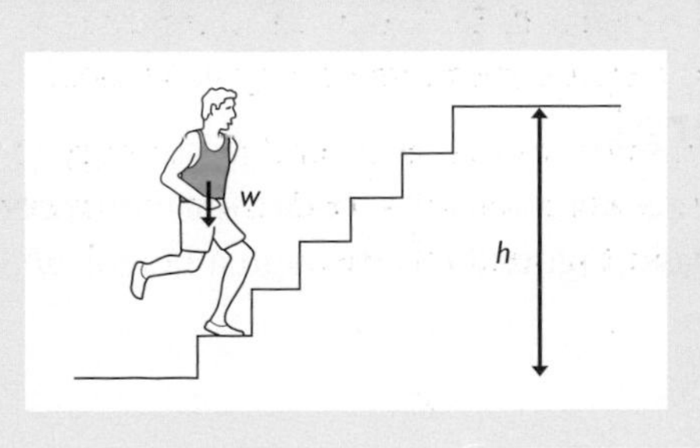

- Measure your weight.
- Then time how long you take to run up a flight of stairs as quickly as you can.
- Measure the height of the flight (Figure 28.6) and calculate the work you did.
- Divide by the time to calculate your power.

Figure 28.6 Running upstairs

29 Potential and kinetic energy

Systems that can do work

You can use the energy of many different types of system to do work. For example, a mixture of food and oxygen has energy that you can use to do work. A petrol–air mixture enables a car engine to do work; a coal–air mixture enables the turbines in a power station to do work. You can use moving bodies to do work, or hot bodies. High bodies can do work; so can squashed springs and nuclear fuels. All these systems have energy, and since they are different, it is tempting to give the energy in them all different names. But when you study energy in detail, you find that there are really only two different ways of storing energy, as *potential* energy or as *kinetic* energy.

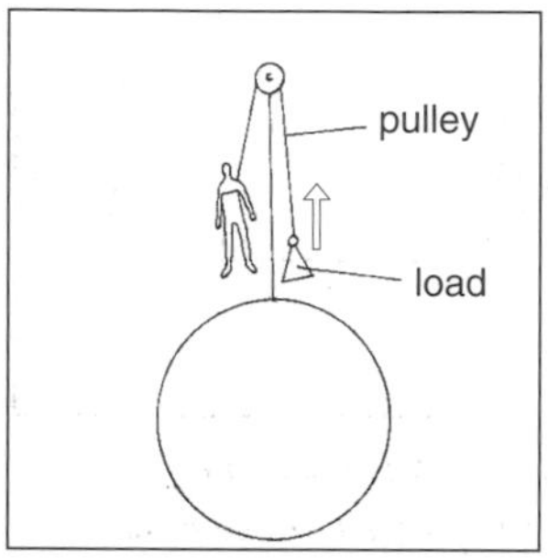

Figure 29.1 As you fall, you do work. The system containing you and the Earth loses energy

Gravitational potential energy

Think about yourself and the Earth as a system. There are gravitational forces between you and the Earth. If you are separated from the Earth, the Earth pulls you down again, and this can do work, for instance by raising a load (Figure 29.1).

If, on the other hand, the load raises you, it does work on you, separating you from the Earth and increasing the energy of your system (Figure 29.2).

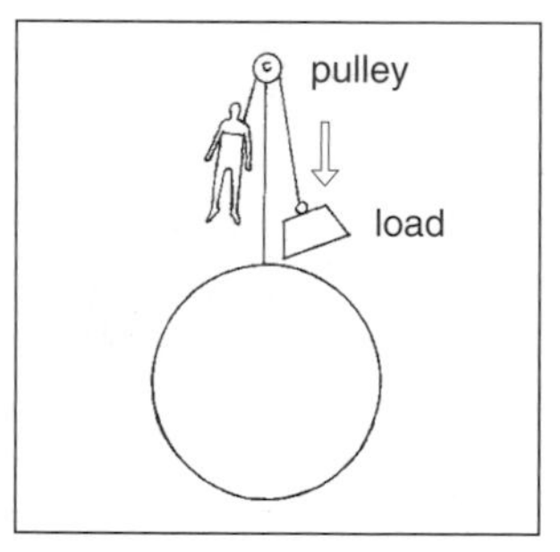

Figure 29.2 The falling mass does work and raises the energy of you and the Earth

If you are high, you have energy. If the load is high, it has energy. In both cases the energy is the consequence of the *position* of a body on which *forces* act. Energy stored like this is called *potential* energy. In this case the forces are gravitational, so the energy is **gravitational potential energy**.

Electromagnetic and nuclear potential energy

Potential energy is also associated with electromagnetic and nuclear forces. If you stretch a spring, you distort the electrostatic bonds between the atoms, and the spring has electromagnetic potential energy, though we often refer to this as elastic potential energy.

Electromagnetic potential energy depends on the arrangement of charges. It is responsible for the energy stored in batteries, capacitors, two magnets squashed together, food plus oxygen, fuel plus oxygen, squashed-up springs and energy released in chemical reactions.

Nuclear fuels have potential energy as a consequence of the strong and weak forces within the nucleus.

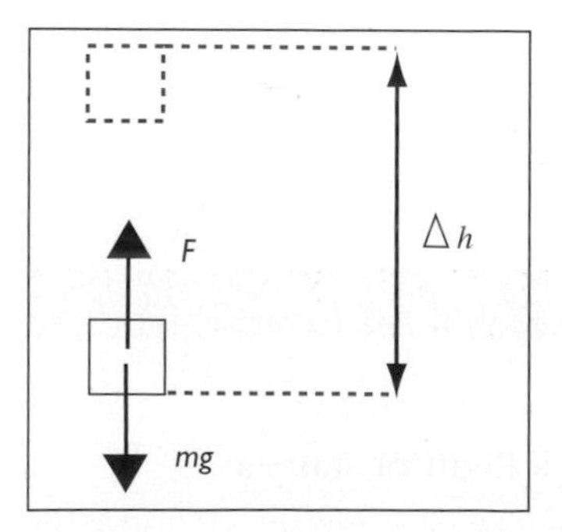

Figure 29.3 Raising a mass through a distance Δh

Calculating potential energy

The force needed to raise a mass m in a gravitational field g at constant speed is mg. The work done in raising this mass (Figure 29.3) through a height Δh is

$$\Delta W = \text{force} \times \text{displacement} = mg\Delta h$$

This is the increase in gravitational potential energy.

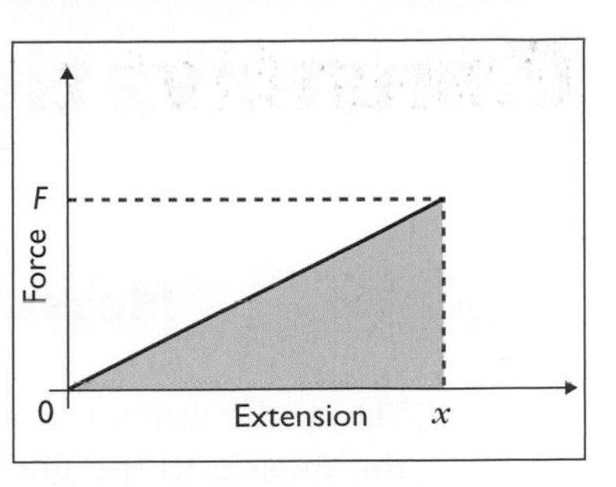

Figure 29.4 Energy stored in spring = area of triangle = $\frac{1}{2}Fx$

Figure 29.4 shows a force–extension graph for a spring. You know from Chapter 28 that the work done is the area under the graph. This is the area of the shaded triangle. This equals the amount of elastic potential energy stored in the spring.

Kinetic energy

You need to do work on a body to get it moving. When it is moving, it has energy – known as **kinetic energy** – and you can use this energy to do work.

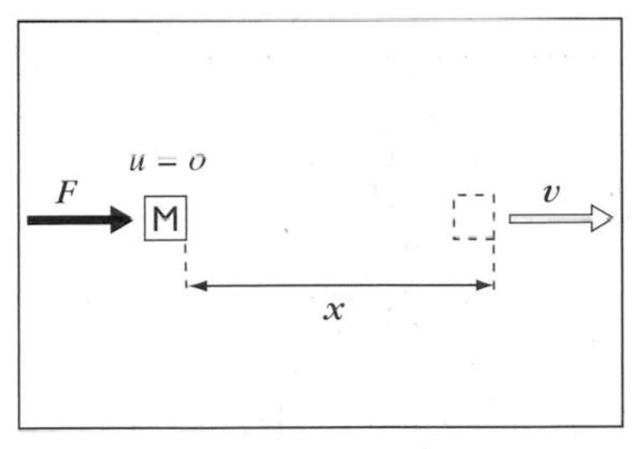

Figure 29.5 Work done in accelerating a mass appears as kinetic energy

You can calculate how much kinetic energy a body has from the amount of work you need to get it moving. When a constant resultant force F accelerates a body of mass m from rest to a velocity v, the work done appears as kinetic energy of the moving body (Figure 29.5). If the acceleration occurs over a distance x then

$$\text{work done} = Fx$$

Since $F = ma$,

$$Fx = max$$

But $v^2 = u^2 + 2ax$, and since $u = 0$, $ax = \frac{1}{2}v^2$,

$$\text{work done} = Fx = max = m(\tfrac{1}{2}v^2) = \tfrac{1}{2}mv^2$$

$$\text{kinetic energy} = \tfrac{1}{2}mv^2$$

Measuring the energy stored in a rubber band

- Take measurements and plot a force–extension graph for the rubber band.
- Use the band to catapult the glider along the air track (Figure 29.6). Measure the extension and use the light gate and card to measure the speed of the glider.
- Use the graph to calculate the energy released by the band and compare this with the kinetic energy of the glider.
- Repeat for a range of extensions.

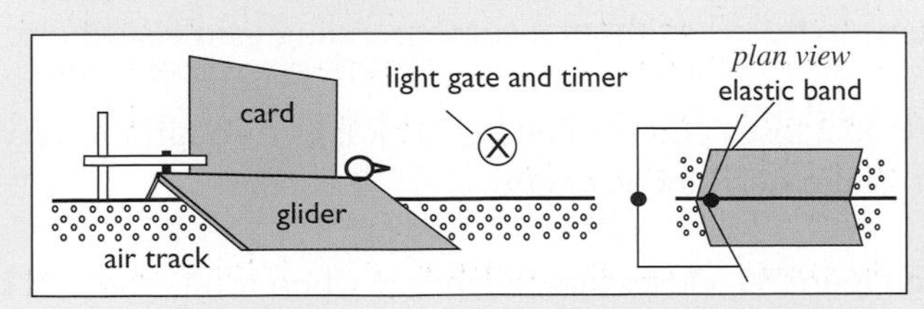

Figure 29.6 Catapulting an air track glider using an elastic band

30 Conservation of energy

Measuring potential energy and kinetic energy

- Set up the apparatus as in Figure 30.1. As the mass falls and loses potential energy, both the mass and the glider gain kinetic energy.
- Calculate the potential energy lost by the mass in pulling the glider to the light gate. Use the light gate and card to measure the speed of the mass and glider and calculate the kinetic energy they gain.
- Compare the potential energy lost with the kinetic energy gained.
- Repeat using different masses and release heights.

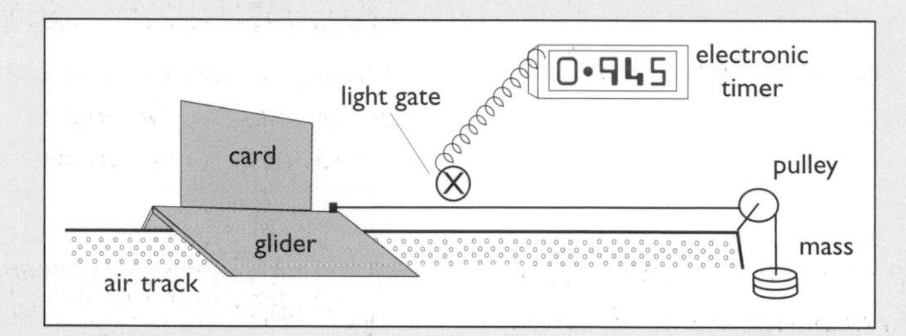

Figure 30.1 Using a falling mass to accelerate an air track glider

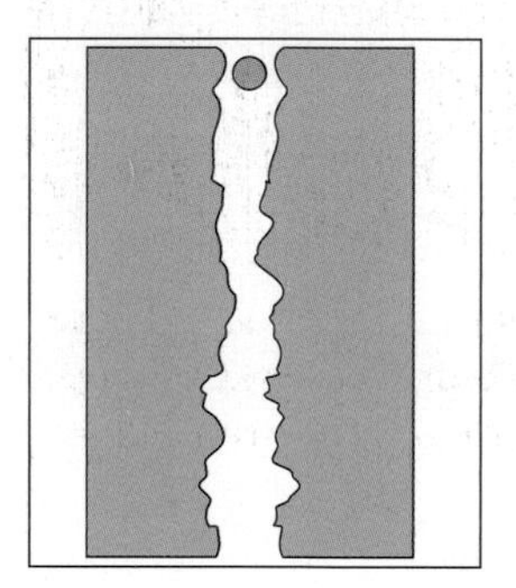

Figure 30.2 The ball gains very little kinetic energy as it falls through a rough tube

Gain and loss of energy

If the air track and pulley in Figure 30.1 are very smooth, the kinetic energy gained by the glider and mass is very nearly equal to the potential energy lost by the mass falling. Of course the connecting string and pulley gain some energy. If you calculate their kinetic energy as well, then you find that the potential energy lost by the mass is even more closely equal to the kinetic energy of the moving parts. In experiments like this, where there is an exchange between potential energy and kinetic energy, you seem to end up with the same amount of energy at the end as you had in the beginning.

Internal energy

It is very easy, however, to think of situations in which potential energy or kinetic energy seems to be lost without a corresponding gain elsewhere.

The mass in Figure 30.2 loses potential energy as it falls through the very rough tube but does not gain much kinetic energy.

Figure 30.3 The mud ball loses kinetic energy as it hits the wall

The ball of mud in Figure 30.3 loses kinetic energy when it hits the wall but nothing seems to gain potential energy.

Observations like these do not seem to fit in with experiments that seem to show that, whenever there is a loss of energy somewhere, there is a gain elsewhere. These observations were a puzzle for scientists for a long time until they noticed that, in such situations where there is a loss of both potential and kinetic energy, the temperature of the bodies involved always increases.

The molecules in bodies are in a continuous state of random motion. They have random kinetic energy. The molecules have electromagnetic forces between them and when they move relative to one another these are stores of random potential energy.

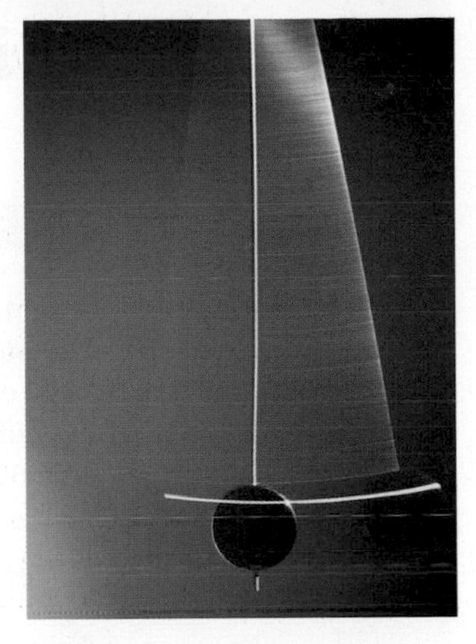

Figure 30.4 The pendulum loses energy to the surrounding air, increasing the air's internal energy

When the ball runs down the rough tube, its potential energy is used to give its own molecules and those of the tube more random kinetic and potential energy. You notice this by the increase in temperature of the ball and the tube. Similarly the temperatures of the mud ball and wall both rise when the ball hits the wall.

The random kinetic and potential energy of the molecules of a body are known as **internal energy**. Internal energy is sometimes referred to as *thermal energy* or, improperly, as heat. You will find out in *Electricity and Thermal Physics* how to calculate the internal energy gained by a body.

Figures 30.4 and 30.5 show situations in which systems increase the internal energy of their surroundings.

Conservation of energy

Careful measurements with many experiments, together with much indirect evidence, has led physicists to state the principle of **conservation of energy**. This states that, in any system isolated from its surroundings, the total amount of energy in that system remains constant. The energy may move about or be rearranged within that system, but there is no increase or decrease; energy is neither created nor destroyed.

Figure 30.5 As the car loses kinetic energy, it raises the internal energy of its brakes and the surroundings

Efficiency

When you walk about or run upstairs, you do work. To do this work you need energy released as your food combines with oxygen from the air. The energy you need is always more than the work you do. No energy is lost; you don't put out less energy than you take in. It is just that you don't use all the energy you take in to do work moving about or raising yourself up.

This is true for almost any process that any machine, engine or mechanical device undertakes. The useful output is never greater than the input, and it is almost always less than the input. For efficient devices, less of the input is wasted and a bigger proportion is useful output.

We define **efficiency** as the proportion of the work or energy input that comes out usefully:

$$\text{efficiency} = \frac{\text{useful output}}{\text{input}}$$

The efficiency of systems varies over a wide range. A manual car gear box is well over 95 per cent efficient; a car engine is less than 40 per cent efficient. You can read more about engine efficiency in *Electricity and Thermal Physics*.

31 Elastic and inelastic collisions

Investigating collisions

- Use spring buffers on the gliders as in Figure 31.1. Send them towards each other so that they collide gently and bounce off each other.
- Find the total kinetic energy before the collision and afterwards. Repeat for a range of speeds.
- Replace the spring buffers with plasticine so that the gliders stick together after colliding.
- Again determine the kinetic energy before and after the collision.

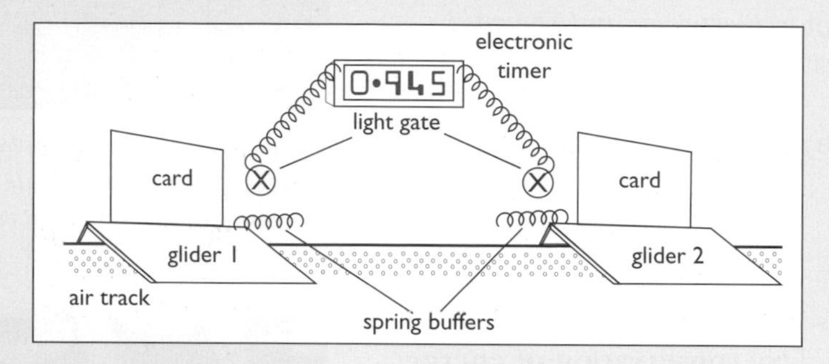

Figure 31.1 Two gliders on an air track before collision

Figure 31.2 An inelastic collision

Types of collision

You know that, in all collisions in systems which are isolated from their surroundings, both momentum and energy are conserved. There are some collisions in which kinetic energy seems to be conserved, and some in which it clearly is not.

If gliders on an air track have very springy buffers, there is no loss of kinetic energy when they collide. These are **elastic collisions**. If the buffers are not springy, for instance if they are made of plasticine, then there are losses of kinetic energy when the gliders collide. These are **inelastic collisions**. Very few collisions between bodies are completely elastic, since there is a general tendency for kinetic energy to be lost as bodies interact and for this to increase the amount of internal energy. A car crash (Figure 31.2) is an inelastic collision.

This trend, from the ordered kinetic and potential energy of bodies to the disordered internal energy inside bodies, is part of the general trend to disorder in the universe. There is more about this subject in *Electricity and Thermal Physics*.

Collisions between the molecules of gases are, on average, elastic. If this were not so, the molecules would gradually slow down and sink to the bottom of the container they were in.

The simple pendulum

In a freely swinging pendulum (Figure 31.3) there is a constant interchange between gravitational potential energy and kinetic energy and back. At the centre the bob is at its lowest point; it has minimum gravitational potential energy and maximum kinetic energy. It is at this point that it is moving at its maximum speed. At the extreme positions the bob is momentarily at rest; it has zero kinetic energy and maximum gravitational potential energy. The effects of friction on a heavy bob are very small over a single oscillation, but the small frictional forces have an effect in the long term. Energy gradually leaks from the pendulum and the internal energy of the surroundings increases.

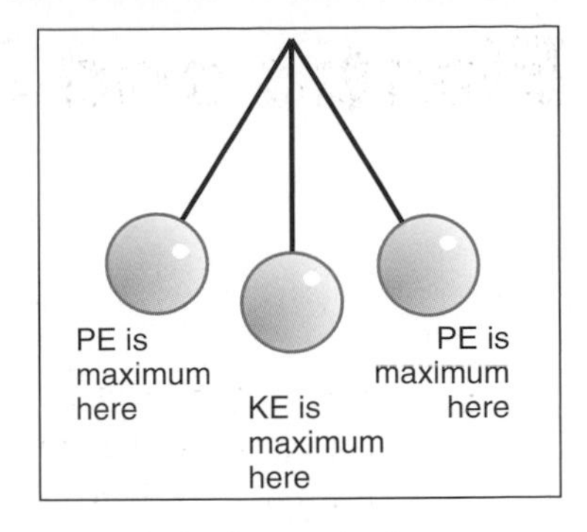

Figure 31.3 In a swinging pendulum, there is a continual interchange between KE and PE

Using a ballistic pendulum to measure the speed of an air rifle pellet

- Find the mass of a pellet and the mass of the plasticine bob. Fire the pellet horizontally so that it embeds in the suspended plasticine (Figure 31.4).
- Record the maximum height h that the bob rises during its swing (Figure 31.5). Calculate the gravitational potential energy the bob and pellet gain as they rise to the highest point. This is equal to the kinetic energy of bob and pellet after the impact. From this calculate the velocity of the bob and pellet just after impact, and their momentum.
- Use the law of conservation of momentum to obtain the initial momentum of the pellet and calculate the initial velocity of the pellet.
- How does the initial kinetic energy of the pellet compare with the kinetic energy of the bob and pellet after impact? Account for any difference you observe.

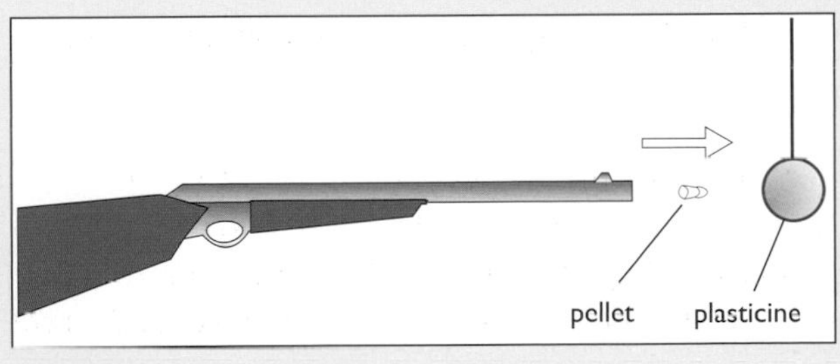

Figure 31.4 The pellet is about to enter the suspended plasticine

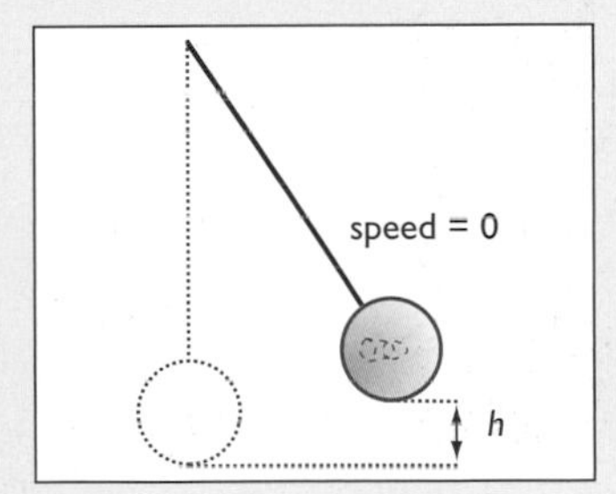

Figure 31.5 What is the maximum height of the pellet and plasticine after impact?

When the pellet from the air rifle hits the plasticine pendulum bob, the collision is inelastic. Most of the pellet's energy is used to change the shape of the plasticine and thereby increase its internal energy. Modern cars are designed with a crumple zone at each end (Figure 31.2) which will change shape during impact and absorb the kinetic energy of the vehicle.

Radioactivity

All objects consist of matter and that matter is made up of atoms. Radioactivity is the spontaneous disintegration of the atom's nucleus, with the consequent emission of particles and energy from the atom. Radioactivity affects us every day. It can be man-made (such as that from medical isotopes), but most of it occurs naturally in the environment around us.

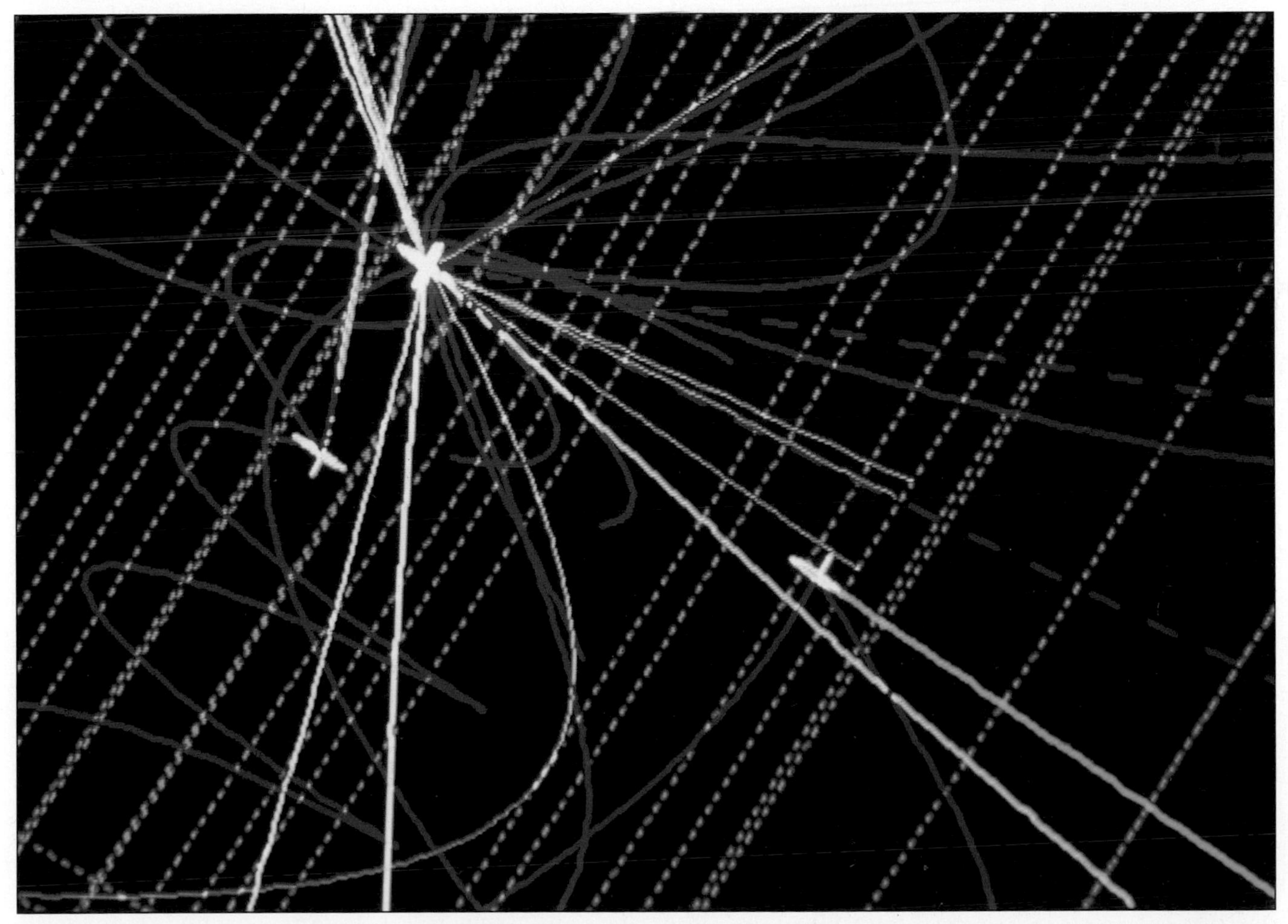

The paths of matter and antimatter (an electron and a positron) in a bubble chamber. The blue lines represent part of the cylindrical detector.

32 Inside the atom

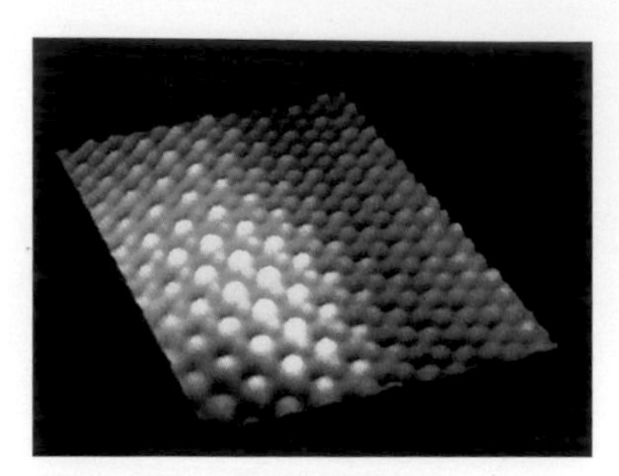

Figure 32.1 Field emission microscope image showing individual atoms of palladium.

Matter is made up of particles

Figure 32.1 shows a field emission microscope image of the surface of a metal. This gives strong evidence that the metal is made of a large number of identical particles fitted together in a regular way.

Atoms have parts

Since Greek and Roman times, people have discussed whether it is possible to cut matter into smaller and smaller pieces, or whether it consists of indivisible particles called atoms. The first clear evidence for the existence of atoms came from chemistry early in the nineteenth century. But final confirmation had to wait until a detailed study of radioactivity early in the twentieth century. Radioactivity gave evidence not only for atoms, but also that atoms themselves are made of smaller particles.

Protons, neutrons and electrons

The three principal parts of the atom are the proton, neutron and electron. The electron was the first to be investigated separately. As a result of J.J. Thomson's work in 1897 and Robert Millikan's in 1909, the electron was known to be part of all atoms; its mass and its negative charge were known. The proton has an equal and opposite charge to the electron. Some properties of the proton were measured before 1900, but the place of the proton in the atom was not established until Ernest Rutherford's work in 1914. The mass of the proton is about 1800 times the mass of an electron. The neutron is an uncharged particle with a mass slightly greater than that of the proton. The existence of the neutron was predicted in 1920 by William Draper Harkins, but it was not until 1932 that James Chadwick actually demonstrated its existence. Table 32.1 summarises these properties of protons, neutrons and electrons.

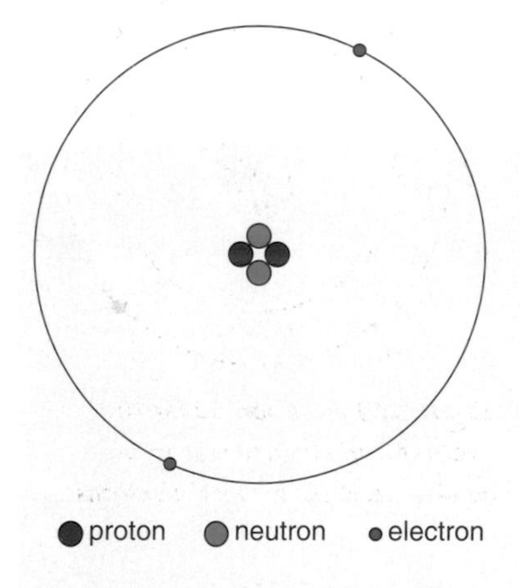

Figure 32.2 The nucleus consists of neutrons and protons. Electrons orbit the nucleus.

Table 32.1 *Properties of the proton, neutron and electron*

Property	Proton	Neutron	Electron
mass	1.7×10^{-27} kg 1 u (atomic mass unit)	1.7×10^{-27} kg 1 u	9.1×10^{-31} kg 1/1800 u
charge	$+1.6 \times 10^{-19}$ C	0	-1.6×10^{-19} C
description	a nucleon; part of the nucleus	a nucleon; part of the nucleus	in a cloud around the nucleus

The nucleus

You can read in the next chapter how physicists discovered that most of the mass of the atom is concentrated in the tiny **nucleus** in the centre of the atom. The rest of the atom is mostly empty space.

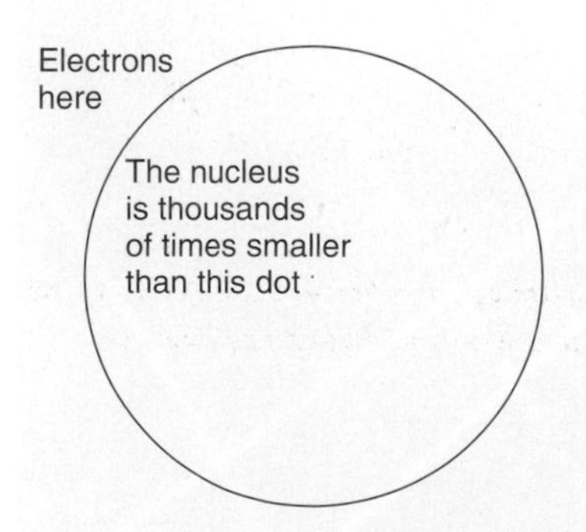

Figure 32.3 A typical diameter of an atom is about 10^{-10} m. A typical diameter of a nucleus is about 10^{-15} m

The nucleus consists of protons and neutrons, relatively massive particles, with the low-mass electrons orbiting in a fuzzy cloud around the nucleus. Figure 32.2 shows an artist's impression of a helium atom. All parts of the atom are magnified to make them visible. In a real helium atom, the particles are tiny compared with the space between them. The nucleus is tens of thousands of times smaller than the atom. Figure 32.3 gives you an idea of the real scale of a helium atom.

Describing atoms

An atom's **proton number**, symbol Z, is the number of protons in an atom. In a neutral atom, the number of protons in the nucleus is equal to the number of electrons round the outside. All the atoms of a particular element have the same proton number, so the proton number of an atom identifies which element it is. The proton number is sometimes called the **atomic number**. Hydrogen (Figure 32.4) has a proton number of 1.

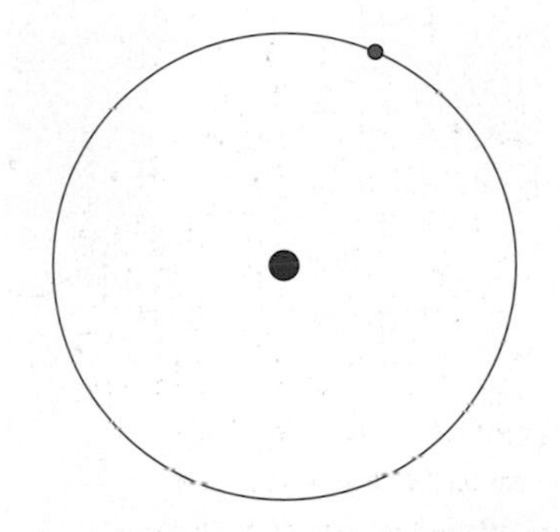

Figure 32.4 This hydrogen atom has one proton and one electron

The proton number of carbon is 6; a neutral carbon atom has six protons and six electrons (Figure 32.5).

Protons and neutrons are **nucleons**. The **nucleon number**, symbol A, of an atom is the total number of protons and neutrons in the nucleus of that atom. The protons and neutrons are the most massive particles in an atom, and they both have approximately the same mass. So the nucleon number gives an approximate indication of the mass of an atom. For this reason, the nucleon number is sometimes called the **mass number**. Both the nucleon number and the proton number are integers (whole numbers).

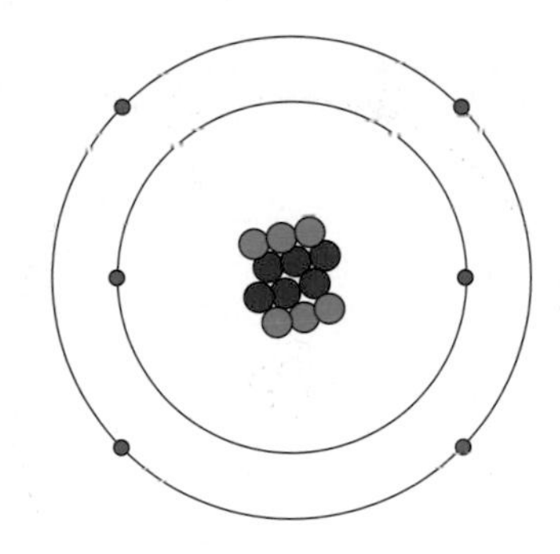

Figure 32.5 Carbon–12 has six protons and six neutrons in the nucleus and six orbiting electrons

To find the number, N, of neutrons in an atom, subtract the proton number from the nucleon number. For example, gold (symbol Au) has a proton number of 79 and a mass number of 197. So for gold:

$$N = A - Z = 197 - 79 = 118$$

$^{197}_{79}Au$ is a shorthand way of indicating the proton and nucleon numbers. So an atom of $^{197}_{79}Au$ has 79 protons, 118 neutrons and 79 electrons.

Isotopes

A nucleus with a stated proton number and neutron number is called a nuclide. An **isotope** of a nuclide is another nuclide with the same proton number but a different neutron number. $^{12}_{6}C$ and $^{14}_{6}C$ are both isotopes of carbon. They have the same number of protons, but $^{14}_{6}C$ (Figure 32.6) has two more neutrons than $^{12}_{6}C$. The proton number determines the arrangement of the electrons around the nucleus, and so decides the chemical properties of an atom. So all isotopes of a nuclide have the same chemical properties because they have the same proton number. They have different atomic masses, and so have different densities.

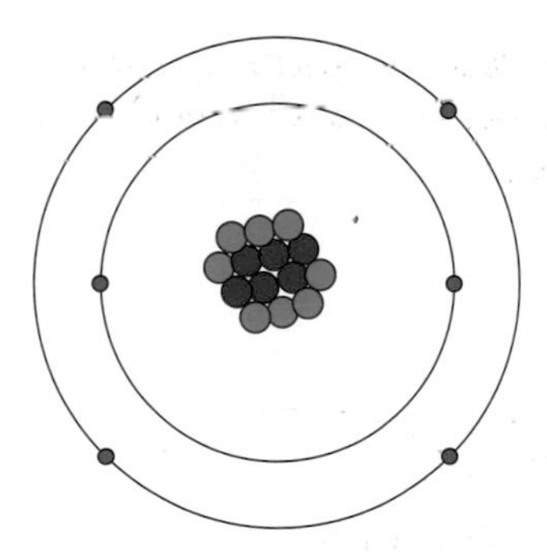

Figure 32.6 Carbon–14 has an extra two neutrons in the nucleus

33 Scattering

Figure 33.1 *Sir Ernest Rutherford*

The nuclear atom

Little was known about the structure of atoms until, in 1911, Ernest Rutherford (Figure 33.1) asked two other researchers, Geiger and Marsden, to carry out an experiment. As you can read in the next chapter, certain atoms give out alpha (α) particles. At the time of **Rutherford's scattering** experiment, these were known to be positive particles that are part of the helium atom. We know now that they consist of two protons and two neutrons.

Alpha (α) particle scattering

- Geiger and Marsden arranged to fire alpha particles at thin gold foil and detect them by a screen, which gave out a flash of light whenever it was hit by an alpha particle (Figure 33.2).
- By counting the flashes, they compared the number of alpha particles passing straight through with the numbers deflected through various angles.

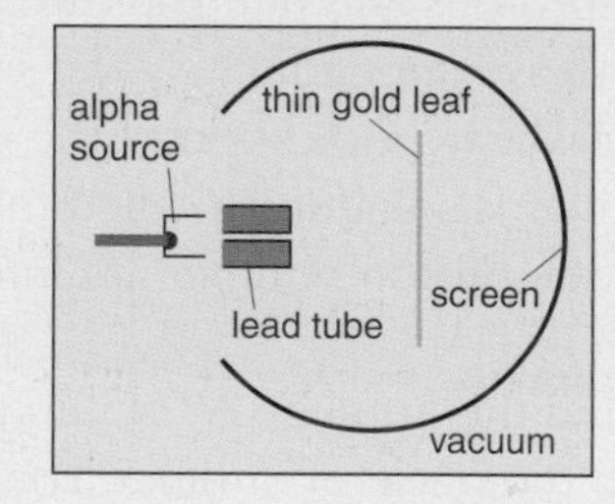

Figure 33.2 Alpha scattering apparatus

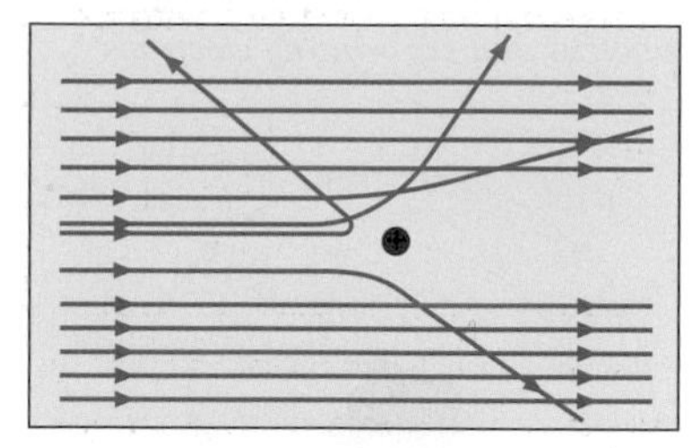

Figure 33.3 About 1 in 8000 alpha particles were deflected through greater than 90°

The results of the alpha particle scattering experiment were surprising. Geiger observed that the vast majority of the alpha particles are deflected very little as they travel through the gold foil, but a tiny minority are deflected through large angles. Figure 33.3 gives a representation of these observations.

Rutherford knew that alpha particles had masses equal to those of light atoms. So whatever deflected the alpha particles would need to have more mass than a light atom. He knew that the negative charges, the electrons, had too little mass to deflect the alpha particles; so the deflection must be due to the positive charges in the atom. These positive charges must have comparatively large distances between them to account for the fact that most alpha particles are undeflected.

Rutherford found the solution to these puzzles in the nuclear atom. He suggested that most of the mass of the atom is concentrated in a tiny positively charged centre called the nucleus. We now know that the nucleus contains both the protons and neutrons, and that the electrons form a cloud in the 'space' of the atom.

Atomic and nuclear sizes

Figure 33.4 Thin gold foil

Thin gold foil (Figure 33.4) can be typically 1/1000th mm (= 10^{-6} m = 1 μm) thick. At this thickness, the foil is about 3000 atoms thick – the diameter of a gold atom is about 0.3×10^{-9} m (= 0.3 nm). The nucleus is more than 10 000 times smaller still – less than 10^{-14} m in diameter. If the gold atom were the same size as this page, the nucleus would be 100 times smaller than a single full stop.

Quarks

For some time, physicists thought that protons, neutrons and electrons alone were sufficient to explain the structure of matter, but measurements of radioactive decay produce evidence of other sub-atomic particles. You can read about these particles in the Particle Physics topic section of *Electricity and Thermal Physics*.

Nowadays, physicists think that all sub-atomic particles are themselves made up of yet smaller particles called **quarks** (which rhymes with either 'pork' or 'park').

Evidence for *quarks* comes from firing electrons at protons. You might think that electrons would stick to protons because of the electrostatic attraction. In fact, if they are fired at high speeds they bounce off. This shows that there are other forces between them. In this case, it is the weak nuclear force. (The weak nuclear force is responsible for the beta decay that you will learn about in Chapter 35.)

The pattern of the rebounds gives information about the distribution of matter within the proton. If the proton is uniform, the field around the proton should be uniform and the electrons should scatter uniformly and elastically. Low energy electrons do scatter in this way (Figure 33.5a) and the protons recoil as predicted. In these elastic collisions kinetic energy is conserved, as you learned in Chapter 31.

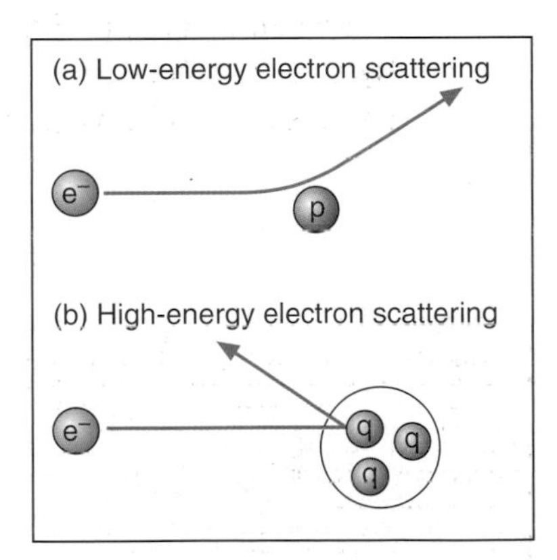

Figure 33.5 Deep inelastic scattering reveals that the proton has structure

Above a particular energy, the protons deflect some electrons through large angles. This is like the nucleus scattering alpha particles. It shows that the proton is not uniform, but has an internal structure. In these sort of collisions, kinetic energy is lost. Where has it gone to? The answer is that it has increased the internal energy of the proton – it has given energy to the particles inside the proton.

This is **deep inelastic scattering**. This scattering is *deep*; it probes deep into the structure of matter. It is also *inelastic*, because sometimes the electron loses kinetic energy (Figure 33.5b).

If protons can do this to electrons, it suggests that the charge in the protons is not uniform but split between even smaller charged particles, the quarks.

34 Alpha radiation

Conducting air

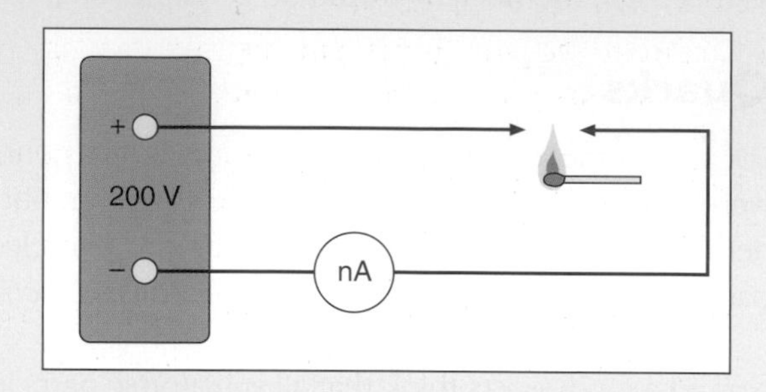

Figure 34.1 Measuring current through an air gap

- Set up the circuit of Figure 34.1. The high-voltage power supply pushes charge around the circuit, and the nanoammeter measures the current that flows. With no match present, check that no current flows through the air gap.
- Now hold a match flame underneath the air gap and observe the meter.
- Then bring an americium radioactive source to the air gap and observe the meter.

Ionising air

When you hold a flame under the air gap in the circuit of Figure 34.1, the meter indicates that a current is flowing. The flame is causing the air to conduct. When you hold the radioactive source near the gap, the same thing happens and the air conducts.

The flame **ionises** the air. It gives atoms in the air sufficient energy to release electrons. This produces free negatively charged electrons and free positively charged ions, which conduct a current. The radiation from the americium also ionises the air; so it is called ionising radiation. The radiation released by the americium is called **alpha (α) radiation**.

Investigating alpha (α) radiation

- The ionisation chamber is an enclosed air gap that you can use to investigate radiation. Any radiation that gets in to the chamber and ionises the air enables a current to flow and be detected by the meter.
- Check that the americium source produces an ionisation current when you hold it near the gauze (Figure 34.2). Then move the source further away from the gauze until the current stops, so that you can find the distance that the alpha radiation can travel through air.
- Next hold the source close to the gauze. Put thin pieces of paper between the source and the gauze to investigate the distance that alpha radiation can travel through paper.

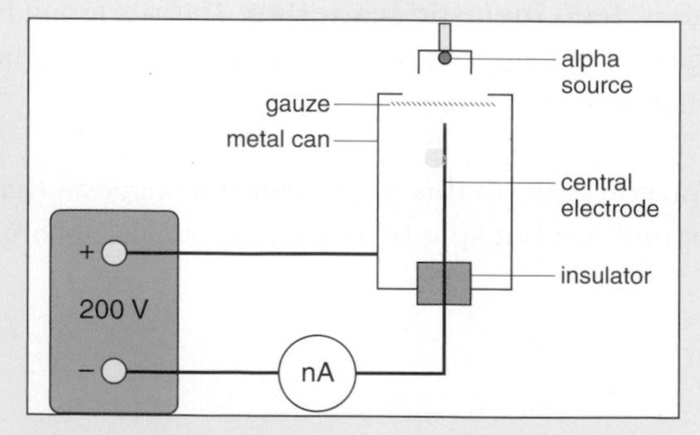

Figure 34.2 Detecting the ions produced by alpha radiation

The range of alpha radiation

Alpha radiation is heavily ionising – it produces many ions per millimetre along its path. But the range of alpha radiation is limited. As Figure 34.3 shows, it is stopped by 5 – 6 cm of air, or by thin paper. Figure 34.4 shows a smoke detector which is triggered by smoke particles stopping alpha radiation.

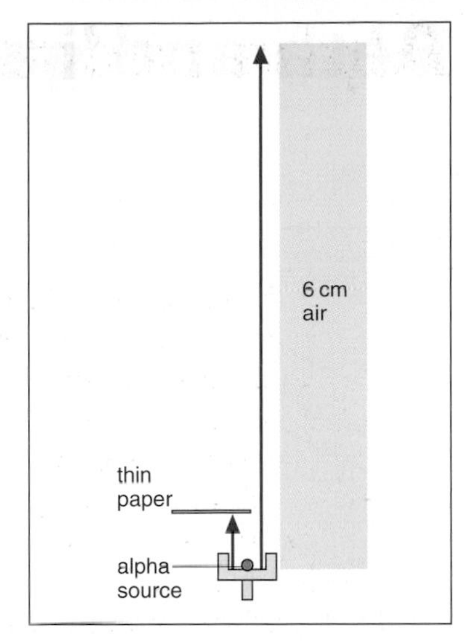

Figure 34.3 Alpha particles are stopped by thin paper or 6 cm of air

The mechanism of alpha decay

You read in Chapter 33 that alpha particles consist of two protons and two neutrons. This is the same as the nucleus of the helium-4 nuclide ($^{4}_{2}He$).

When an atom decays by emitting an alpha particle, it emits two protons and two neutrons (Figure 34.5)

Americium decays by emitting an alpha particle. The proton number decreases by 2 and the nucleon number decreases by 4. The atom becomes neptunium; it is no longer americium. This equation shows the decay:

$$^{241}_{95}Am \rightarrow ^{237}_{93}Np + ^{4}_{2}\alpha$$

Neptunium decays by a further alpha decay to protactinium-233:

$$^{237}_{93}Np \rightarrow ^{233}_{91}Pa + ^{4}_{2}\alpha$$

When, at the end of its path, an alpha particle stops, it picks up two electrons and becomes a helium atom. Alpha decay within the Earth is responsible for the presence of helium in natural gas deposits.

Figure 34.4 Smoke detectors contain an americium source. Smoke particles stop the alpha radiation and trigger the alarm

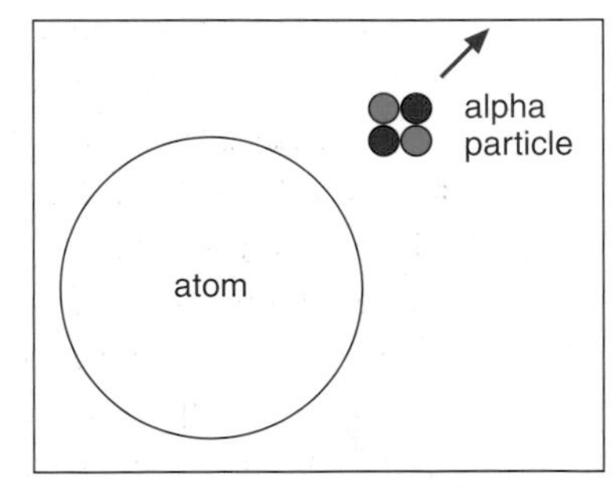

Figure 34.5 In alpha decay, a nucleus emits two protons and two neutrons

35 Beta and gamma radiations

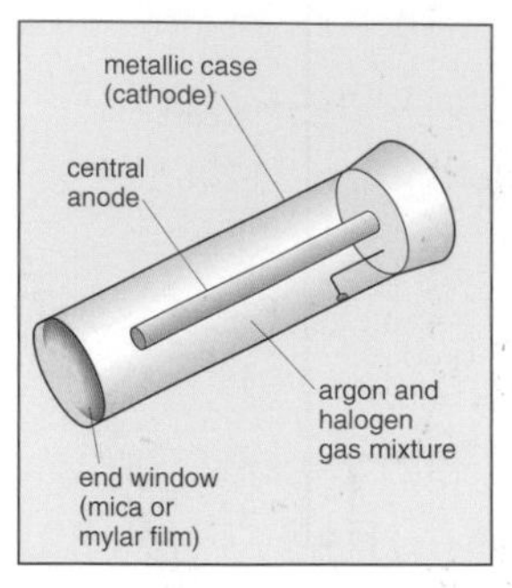

Figure 35.1 A Geiger–Müller tube

The Geiger–Müller tube

The ionisation chamber needs a large number of ions to be produced for a measurable current. A Geiger–Müller (GM) tube (Figure 35.1) will detect each ionising event inside the tube, whether the event involves a single pair of ions, or many ions. For each ionising event, a pulse of current passes through the tube and is recorded by the attached counter. GM tubes are poor detectors of alpha radiation, because even the thin end window stops many alpha particles. But GM tubes can detect more penetrating, but less ionising, radiations that ionisation chambers are not able to detect.

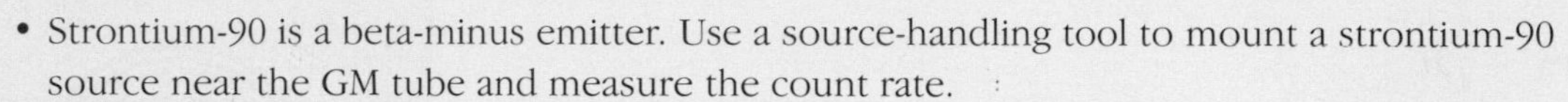

Investigating beta-minus (β^-) and gamma (γ) radiations

Investigating beta-minus radiation

- Strontium-90 is a beta-minus emitter. Use a source-handling tool to mount a strontium-90 source near the GM tube and measure the count rate.
- For a range of distances from the tube, measure the count rate. Plot a graph of count rate against distance.
- Fix the beta-minus source 3 cm from the GM tube and measure the count rate (Figure 35.2).
- Insert a piece of paper between the source and the tube, and measure the new rate.
- Then insert a series of thin pieces of aluminium between the source and the tube. Plot a graph of count rate against number of pieces of aluminium.

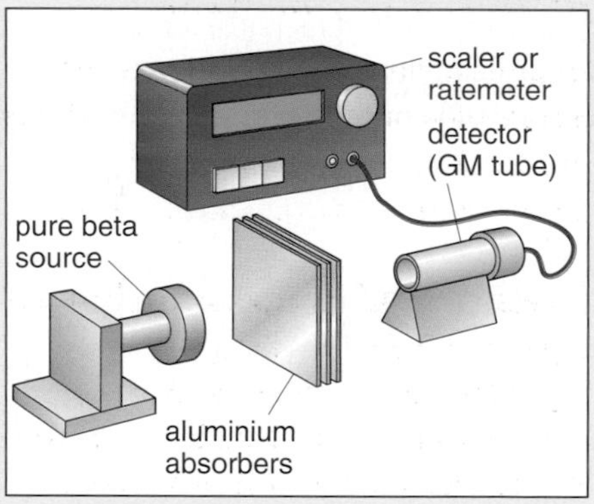

scaler or ratemeter
detector (GM tube)
pure beta source
aluminium absorbers

Figure 35.2 Investigating beta radiation

Investigating gamma radiation

- Use the arrangement in Figure 35.2 to measure the count rate at different distances from a cobalt-60 gamma source.
- Then use a source-handling tool to mount the source 8 cm from the tube and measure the count rate for a range of thicknesses of lead absorbers between the source and the tube.

Properties of beta-minus (β^-) radiation

Beta-minus (β^-) radiation is much less heavily ionising than alpha radiation, and so is very difficult to detect with an ionisation chamber. However, beta-minus radiation will travel more than 30 cm through air and through several millimetres of aluminium.

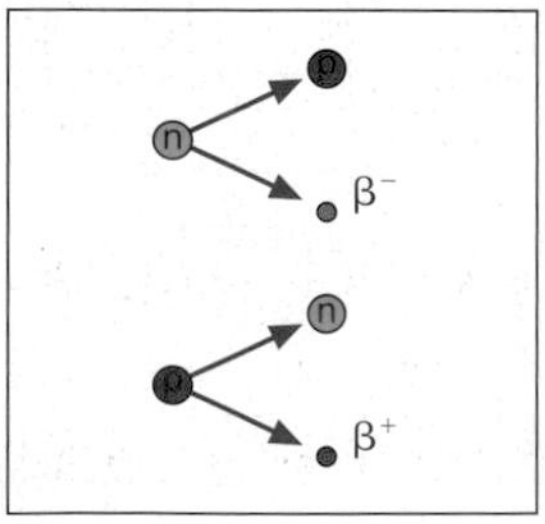

In beta-minus decay, a neutron in the nucleus splits up into a proton plus an electron. The proton stays in the nucleus; the electron is ejected at high speed – it is a beta-minus particle. Beta-minus particles are fast electrons that have been emitted from the nucleus.

Figure 35.3 β^- and β^+ decays

When beta-minus decay occurs, the number of nucleons stays the same. The number of protons goes up by 1, the number of neutrons goes down by 1 and an electron is emitted. Strontium-90 becomes yttrium-90, as the equation shows:

$$^{90}_{38}\mathrm{Sr} \rightarrow {}^{90}_{39}\mathrm{Y} + {}^{0}_{-1}\beta^-$$

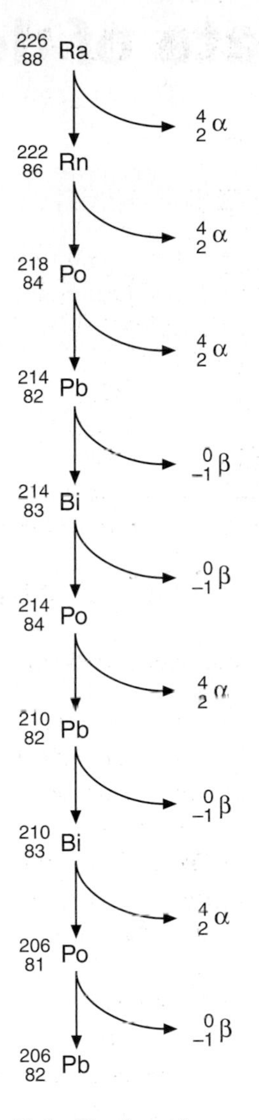

Figure 35.4 The decay from radium-226 to stable lead involves a series of both α and β decays

Beta-plus (β^+) decay

Another type of decay, beta-plus (β^+) decay, occurs rarely in Nature, but more frequently in man-made radionuclides. In beta-plus decay, a proton in the nucleus splits up into a neutron plus a positron. A **positron** is a particle of anti-matter. It has the same mass as an electron, but a positive charge. In beta decay, the neutron remains; the beta-plus particle is ejected at high speed.

When beta-plus decay occurs, the number of nucleons stays the same. The number of neutrons goes up by 1, the number of protons goes down by 1 and a positron is emitted. Figure 35.3 shows β^- and β^+ decays.

Carbon-11 decays by beta-plus decay to boron-11, as the equation shows:

$$^{11}_{6}\mathrm{C} \rightarrow {}^{11}_{5}\mathrm{B} + {}^{0}_{1}\beta^+$$

Gamma (γ) radiation

After emitting alpha or beta radiation, a nucleus may have surplus energy. Often it gives out this energy by emitting electromagnetic radiation. These lumps or *photons* of radiation are called **gamma (γ) rays**. You can read more about photons in *Waves and Our Universe*.

Properties of gamma radiation

The gamma radiation from a source is a stream of electromagnetic photons. The photons ionise when they react drastically with matter along the path, knocking a single electron from an atom and therefore producing a single ion pair. The ejected electron produces further ion pairs as it collides with other atoms. When a gamma photon ionises an atom, the number of photons decreases, but the remaining photons are unchanged. Gamma radiation causes relatively little ionisation per millimetre of its path. Because it interacts so little, it has a large range. High-energy gamma radiation is attenuated (reduced in strength), but not stopped, by several centimetres of lead.

Table 35.1 *Summary of the properties of the different radiations*

Property	Alpha (α)	Beta-minus (β^-)	Beta-plus (β^+)	Gamma (γ)
charge	+2	−1	+1	0
rest mass	4 u	1/1800 u	1/1800 u	0
penetration	5 cm air; thin paper	30 cm air; few mm Al	annihilated on interaction with an electron	long way; keeps going through Pb
nature	helium nucleus	electron	positron	e.m. wave
ionising	heavily	light		a single ion pair on interaction

36 Rate of decay

Measuring background radiation

- Set up a GM tube and counter with no source within 5 metres.
- Measure the counts in one minute, ten times in a row.
- Display your results as a histogram.

Background radiation

Radiation is around us all the time, in the background. It is mainly due to Nature but partly due to human activity. Naturally occurring radio-isotopes – those which exist without human intervention – contribute to this **background radiation**. The biggest source of these is in uranium deposits in the ground, which in turn produce other isotopes that decay. A common naturally occurring radio-isotope is radon gas, which spreads throughout the ground and then into the air we breathe. Next to smoking, it is the biggest cause of lung cancer. The other big natural contributor to background radiation comes from cosmic rays – mostly fast-moving nuclei from outer space. They bombard, and are mostly absorbed by, the atmosphere, sending showers of particles to the Earth's surface.

Nuclear fission is the break up of very large nuclei into medium-mass fragments. It occurs in nuclear reactors and the explosion of nuclear weapons. The large nuclei have far more neutrons than protons. When they split, the fragments have too many neutrons for stability. So the fragments are beta-minus emitters and some also emit gamma rays. Useful radio-nuclides are obtained by chemical processing of discharged nuclear fuel. Other useful radio-nuclides can be produced by bombarding atoms with the immense numbers of neutrons in a reactor. As neutrons are uncharged, they are easily captured by many nuclei. Machines such as cyclotrons can bombard materials with protons, deuterons or heavier nuclei.

Radioactivity produced by human activity is very useful, particularly to medicine, where radioactivity is used in the diagnosis and treatment of illness. (It is responsible for less than 1% of background radiation.) You can read more about this in the medical physics topics in *Electricity and Thermal Physics*.

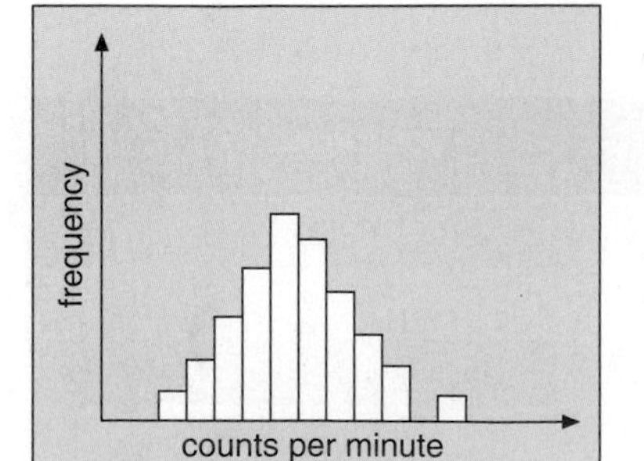

Figure 36.1 A histogram of background counts

Random decay

Figure 36.1 shows a histogram of a large number of one-minute background counts. The fluctuation in count occurs because all radioactive decay is **random** – there is no way of predicting when a particular nucleus will decay. But when there are large numbers involved, there are general statistical patterns.

The number of atoms of a source that decay per second is called the **activity** of the source. If you measure the activity of an istotope over a period of time, you find that the activity will slowly decline. This is because the radioactive decay reduces the number of atoms of that isotope. With fewer atoms remaining, there are fewer left to decay – which is why the activity decreases. The general decline in the level of activity can be modelled by throwing dice, where each die represents a nucleus.

Throwing dice

- Take 100 dice. Throw them all together and remove those which show a six. Stack these in a line.
- Then throw the remaining dice, and again remove those which show a six. Stack these in a line next to the first one. Repeat until all the dice are removed.
- Produce graphs of number remaining for each throw and number removed for each throw.

The decay equation

The throw of one die is, like radioactivity, random. But a pattern emerges if you throw a large number. On average, each die will show a six once in every six throws. You cannot predict which throw will produce a six. But if you throw 100 dice and remove those which show a six, you will remove on average about 16 dice after the first throw. Figure 36.2 shows graphs for the number of dice remaining and the number of dice removed if you continue this process.

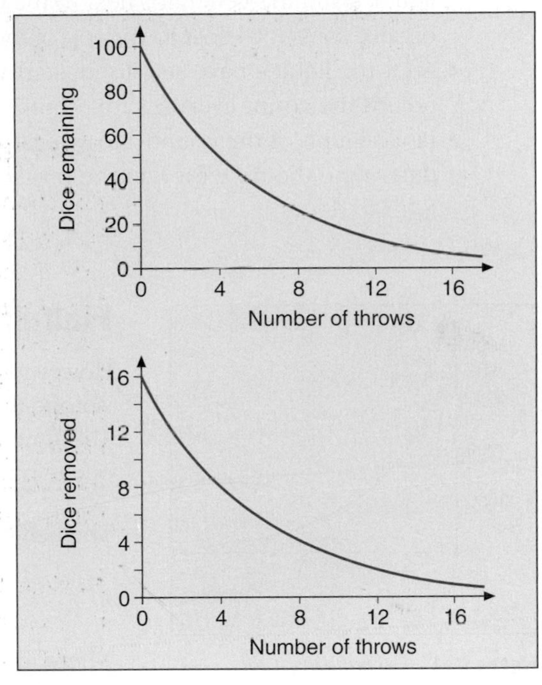

Figure 36.2 Modelling radioactive decay with 100 dice

On any one occasion, the number showing a six, N_6, is proportional to the total number of dice thrown, N:

$$N_6 \propto N \quad \text{or} \quad N_6 = \text{constant} \times N$$

In this case the constant would be 1/6.

Similarly, the activity of a radioactive source is proportional to the total number N of atoms present.

So we can write

$$\text{activity} = \lambda N$$

where N is the number of nuclei present at that instant and λ (the Greek letter *lambda*) is the **decay constant**; λ has units second^{-1}, and represents the proportion of N that decays in 1 s.

If the average rate of decay for 72 nuclei is 12 per second, then activity $= 12\ \text{s}^{-1}$ for $N = 72$. So

$$\text{Activity} = \lambda N$$

$$12\ \text{s}^{-1} = \lambda \times 72$$

$$\lambda = \tfrac{1}{6}\ \text{s}^{-1}$$

Activity is measured in counts per second, called the becquerel (Bq). As the decay proceeds, N is smaller at the start of each successive time interval. So the activity also gets smaller and smaller as the decay proceeds.

The decay constant λ has a different value for each radioactive isotope, because each isotope decays at a different rate. If the decay is rapid, λ is large. If the decay is slow, λ is small. Note that λ is constant for a particular isotope and does not vary during the decay. The rate changes because N changes.

Measuring half-life

- Protactinium-234 is a beta-minus emitter; a compound of it is soluble in organic solvent. It is generated in a water-based solution in the protactinium generator. Shake the generator gently to dissolve the protactinium compound from the water-based liquid, stand the generator next to the GM tube and allow the organic solvent to float to the top of the water (Figure 36.3).
- After the liquids have stabilised, start the counter and record the count every 3 s for 5 min.
- Plot a graph of the count rate against time. From this, determine the time for half the protactinium in the top layer to decay.

radioactive

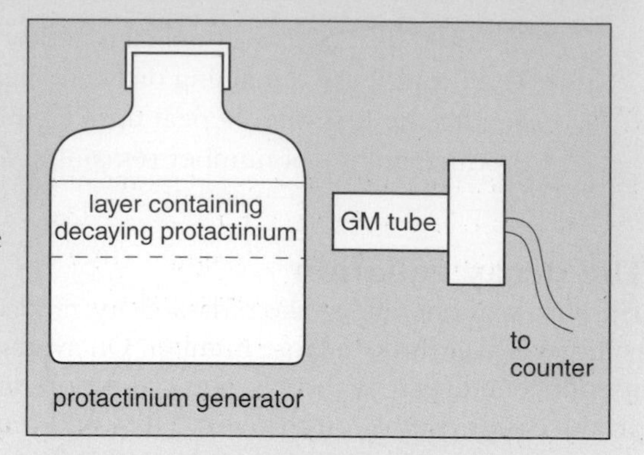

Figure 36.3 Measuring the decay of protactinium dissolved in the top liquid layer

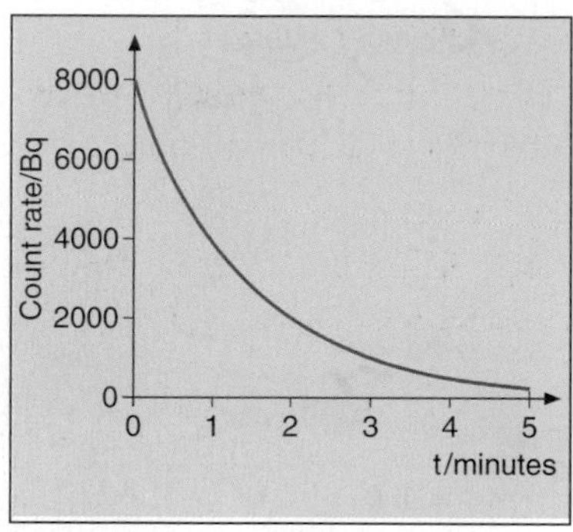

Figure 36.4 The half-life of this isotope is about 1 minute

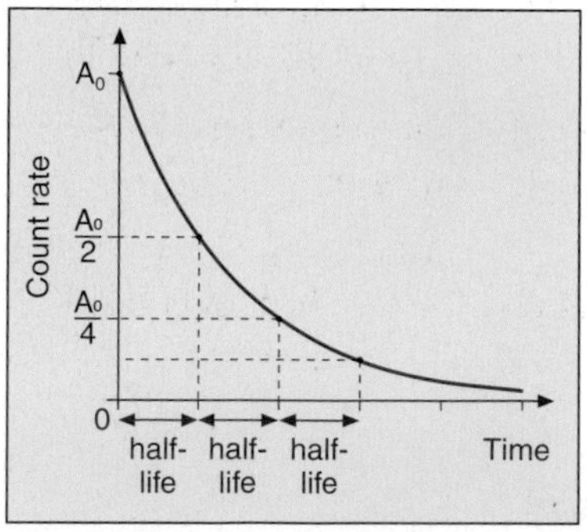

Figure 36.5 The half-life is the average time for the activity of an isotope to halve

Half-life

However large the radioactive sample, experiment shows that for any particular isotope the average time for half of the atoms of one isotope to decay is constant. This time is called the **half-life**. The half-life of the isotope in Figure 36.4 is about one minute.

The half-life $t_{1/2}$ is connected to the decay constant λ by the equation

$$t_{1/2} = \frac{\ln 2}{\lambda}$$

where ln 2 is 0.69.

The half-life of the isotope in Figure 36.4 is about 1 minute = 60 s.

$$\text{Therefore} \quad 60\text{ s} = \frac{0.69}{\lambda}$$

$$\text{and} \quad \lambda = \frac{0.69}{60\text{ s}}$$

$$= 0.012\text{ s}^{-1}$$

As Figure 36.5 shows, the activity will halve from A_0 (the initial activity) to $\frac{A_0}{2}$ (half the initial activity) in one half-life. After a further half-life, the activity will halve again to $\frac{A_0}{4}$ and so on.

Practice questions

Chapter 1

1.1 Vernier callipers are used to obtain the following dimensions of a small rectangular block: length = 74.4 mm; width = 32.7 mm; height = 28.9 mm. Given that the callipers have a zero error of –0.4 mm, what is the correct value of each dimension?

1.2 Describe how you would measure the thickness of the pages in this textbook using a micrometer.

1.3 A metre rule has a resolution of 1 mm. Explain why a student might get an uncertainty of more than 1 mm when using it to measure the height of a room.

1.4 The thickness of a coin is found to be 1.23 mm using a micrometer. If the uncertainty in this reading is 0.05 mm, calculate the percentage uncertainty.

1.5 The resistance of a resistor is quoted as 4.7 kΩ ± 2%. Calculate the range of possible values for this resistance.

Chapter 2

2.1 Describe how you would measure the density of an irregularly shaped solid object that is known to have a greater density than that of water.

2.2 Calculate the mass of (a) air (density 1.3 kg m^{-3}) in a room measuring 4 m × 3 m × 2 m, (b) the Earth (average density 5500 kg m^{-3} and average diameter 12.7 Mm), (c) a cylindrical iron rod (density 8.0 g cm^{-3}, diameter 0.4 cm and length 24 cm).

2.3 A small aluminium cube has a mass of 170 g. The density of aluminium is 2.7 g cm^{-3}. Calculate the length of the sides of the cube.

2.4 A glass cube has sides of length 5 cm. The glass from which the cube is made has a density of 2.5 g cm^{-3}. Calculate the mass of the cube.

2.5 Think about a DVD. Estimate the diameter, the thickness and the mass. Use your figures to calculate a value of the density of the plastic. Compare your answer with the figures on page 5 and comment on your comparison.

2.6 A narrow glass capillary tube contains 10.1 g of mercury in a thread of length 10.5 cm. Given that the density of mercury is 13.6 g cm^{-3}, calculate the average internal diameter of this capillary tube.

Chapter 3

3.1 List the six base quantities that you will be using in your Advanced GCE course. Give the unit of each of these quantities.

3.2 The SI unit of length, the metre, used to be defined in terms of a bar of a standard length (the prototype metre) kept in Paris. State the modern definition of the metre. What are the advantages of using this definition? Are there any disadvantages?

3.3 How many oscillations does a caesium atomic clock make in a day?

3.4 Give three examples of different physical quantities. Use one of your answers to explain how the magnitude of a physical quantity is expressed as the product of a number and a unit.

3.5 How many times bigger is the diameter of the Earth (13 Mm) than that of a human hair (13 μm)?

Chapter 4

4.1 What is a derived quantity?

4.2 Speed, area and volume are examples of derived quantities. Write down their derived units.

4.3 Write down the word equation for density. Use this equation to show that the unit of density is kg m^{-3}.

4.4 Explain what is meant by *homogeneous*. Why must equations be homogeneous if they are to be correct?

4.5 Make up two equations that are homogeneous but still incorrect.

Chapter 5

5.1 A student walks a distance of 3.5 km from home to college. He returns home via the chip shop, covering a distance of 5.5 km. Find the total distance he has walked and his displacement from home at the end of the day.

5.2 Distinguish between a scalar and a vector quantity. Consider the following physical quantities: acceleration, distance, energy, weight, volume, displacement, speed, mass, velocity, force. Which of these are scalars and which are vectors?

5.3 Calculate the average speed in m s^{-1} of (a) a sprinter who completes 100 m in a time of 10 s, (b) a marathon runner who takes 2¼ hours to run 42.5 km.

5.4 Describe and explain how you would use a light-gate and a timer to find the average speed at which a trolley passes a point.

5.5 A cyclist travels to work at an average speed of 3 m s^{-1} and returns home for tea at an average speed of 9 m s^{-1}. Calculate her average speed for the whole journey. [Think carefully … the answer is not 6 m s^{-1}!]

Chapter 6

6.1 A car moving at 3.4 m s^{-1} takes 6.2 s to reach a speed of 25.1 m s^{-1}. Calculate its average acceleration.

6.2 How long will Concorde, flying at a speed of 75 m s^{-1}, take to reach the speed of sound (330 m s^{-1}) if its average acceleration is 5 m s^{-2}?

6.3 Describe how you would use a light-gate and a double interrupter card to find the acceleration of a trolley on a sloping runway. Explain clearly what measurements are taken and how the acceleration is calculated from them.

6.4 What is meant by the rate of doing work?

6.5 Calculate the acceleration of the car in Figure 6.5, page 15.

Chapter 7

7.1 Copy out the following statements, choosing the correct word in each case: The area/gradient of a displacement–time graph is the instantaneous velocity. The area/gradient of a velocity–time is the instantaneous acceleration. The area/gradient of a velocity–time graph is the change in displacement. The area/gradient of an acceleration–time graph is the change in velocity.

7.2 Use the distance–time graph to find: (a) how far the body has moved after 10 s, (b) how long the body takes to travel 8 m, (c) the average speed of the body.

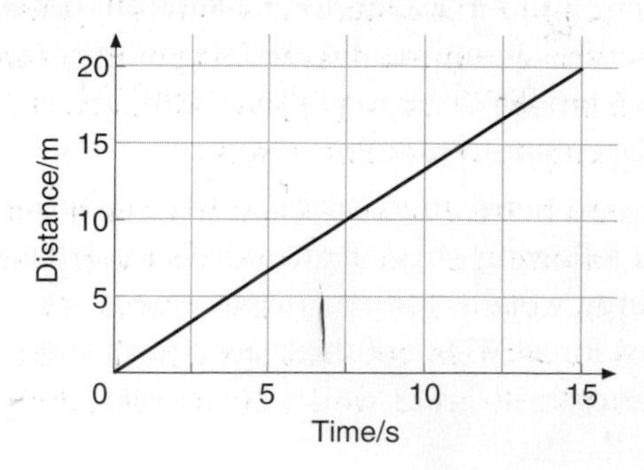

7.3 Look at the velocity-time graph. (a) Describe the motion of the object during the first 5 s, the next 5 s and the last 12 s. (b) Use the graph to calculate the accelerations and distances travelled for each of the three stages. (c) Determine the total distance travelled and the average speed during the whole 22 s journey.

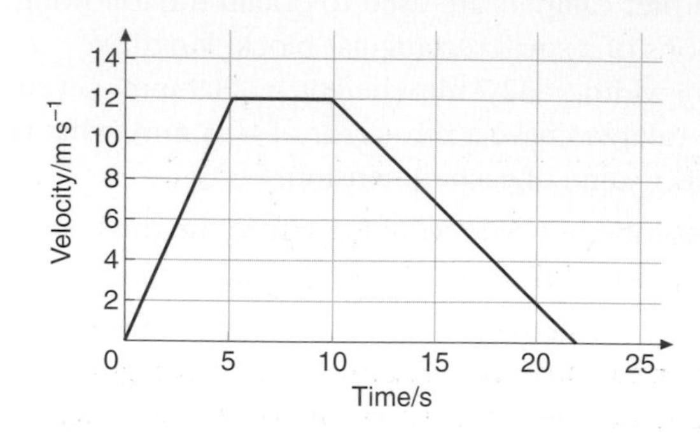

7.4 Sketch the acceleration–time graph corresponding to the velocity–time graph in the previous question.

7.5 An object accelerates from rest with a constant acceleration of 4 m s^{-2}. Sketch the shapes of the following graphs for the first 3 s of its journey, labelling both axes with accurate scales: (a) acceleration against time; (b) velocity against time; (c) displacement against time.

Chapter 8

8.1 Describe how you would obtain the data for a displacement–time graph for a bouncing ball if a motion sensor was not available.

8.2 Sketch a velocity–time graph for a bouncing ball from the moment it is released until its fourth impact with the ground. Take upward as positive. Indicate on your graph two areas that represent the height of the second bounce. How would you find the acceleration of gravity from your graph?

8.3 Sketch a velocity–time graph for a ball that is thrown vertically up into the air from the moment that it leaves the hand until it is caught back at the same horizontal level. Take upward as positive. How does this graph compare to that for a bouncing ball?

8.4 This displacement–time graph is for the motion of a glider between two elastic buffers on an air track.

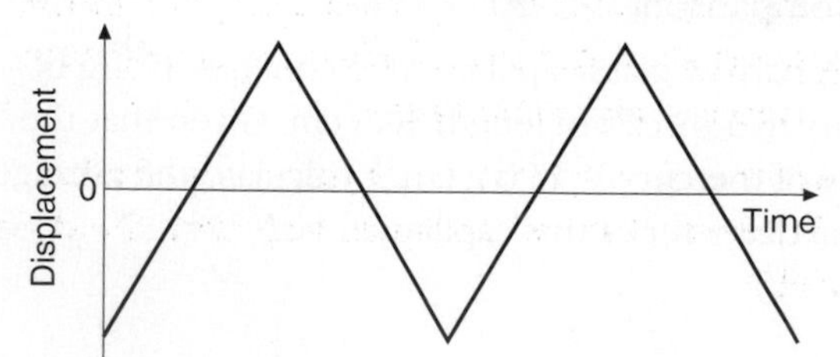

From what point on the track is the displacement being measured? Copy this graph and sketch below it two more graphs, using the same time-scale, showing how the velocity and acceleration of the glider vary with time.

8.5 This displacement–time graph is for the motion of a swinging pendulum.

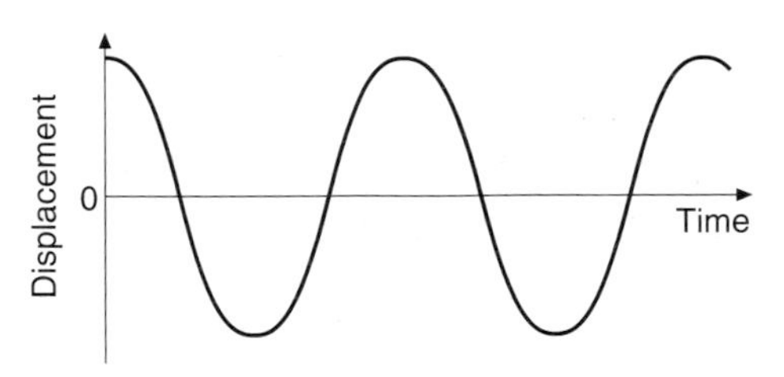

Copy this graph and sketch below it two more graphs, using the same time-scale, showing how the velocity and acceleration of the pendulum vary with time.

Chapter 9

9.1 How long will a car, accelerating at 1.5 m s^{-2} from rest at traffic lights, take to reach the speed limit of 18 m s^{-1} (40 mph)?

9.2 A train moving at 18 m s^{-1} goes through a red light which causes its brakes to be applied automatically. The train stops after 4.5 s. Find its acceleration and the distance it travelled beyond the red light.

9.3 A body with an initial speed of 3.6 m s^{-1} accelerates for 4.5 s at 1.4 m s^{-2}. How far does it travel while accelerating?

9.4 A rocket-powered sledge accelerates at 35 m s^{-2} and takes 1.6 s to pass through a 60 m section of a test track. Find the speed at which it entered the section. What was its speed when it left?

9.5 An electron moving at a speed of 1.0×10^5 m s^{-1} travels 20 cm through an electric field. It leaves in the same direction with a speed of 9.0×10^5 m s^{-1}. Find (a) the acceleration of the electron while it is in the electric field, (b) the time it spends in the electric field.

Chapter 10

10.1 Describe how to determine the acceleration of a freely falling object. Include a labelled diagram of the apparatus and show how to calculate the acceleration from a suitable straight-line graph.

10.2 A stone dropped from rest down a well takes 1.9 s to hit the surface of the water. Calculate the depth of the well.

10.3 How long will it take a pebble dropped from rest from a 30 m high cliff to reach the bottom? Calculate the pebble's speed immediately before it strikes the ground.

10.4 Calculate the speed with which a bullet must be fired vertically in order to reach a height of 200 m. How long does it take for the bullet to reach this height? What is the total distance travelled by the bullet from leaving the gun to returning to the same level? What is the final displacement when it is back to the original level?

10.5 A rocket is launched vertically. For the first 6 s it has an acceleration of 150 m s^{-2}. (a) Calculate its speed at the end of this time, and the distance covered. (b) The rocket continues its flight acted upon only by gravity. Calculate how long it takes to reach the top of its flight and the further distance covered. (c) Sketch the acceleration–time and velocity–time graphs for the upward motion of the rocket.

Chapter 11

11.1 Copy out the following statements, choosing the correct word in each case: A body thrown horizontally from a cliff top takes a/the longer/same/shorter time to reach the bottom as a body dropped vertically. Provided air resistance is large/small, the horizontal/vertical velocity of a projectile is constant while its horizontal/vertical velocity increases at 9.8 m s^{-2}.

11.2 Calculate the time it takes an object to fall from rest through a height of 0.85 m. Repeat the calculation for an object falling through 2 m.

11.3 Bill kicks a football horizontally off the edge of a bench that has a height of 0.85 m. The ball travels a horizontal distance of 6.4 m through the air before it hits the ground. Draw a sketch of this experiment, labelling the distances. By considering only the vertical motion, find how long it takes the football to reach the floor. Then calculate the speed at which it left the bench.

11.4 A bullet is fired horizontally at 400 m s^{-1} from a gun held 2 m above the ground. How long does it take the bullet to reach the (horizontal) ground? Calculate the horizontal distance the bullet travels.

11.5 A dart leaves the thrower's hand horizontally at a height of 1.9 m above the ground. It strikes the board 3 m away at a height of 1.5 m above the ground. Calculate the time taken by the dart to reach the board and the horizontal velocity at which the dart left the thrower's hand.

PRACTICE QUESTIONS

24.5 The diagram shows a pub sign, weight 90 N, attached to one end of a very light uniform rod, the other end of which is hinged to the wall. The length of the rod is 1.5 m. The tension in the wire keeps the rod horizontal.

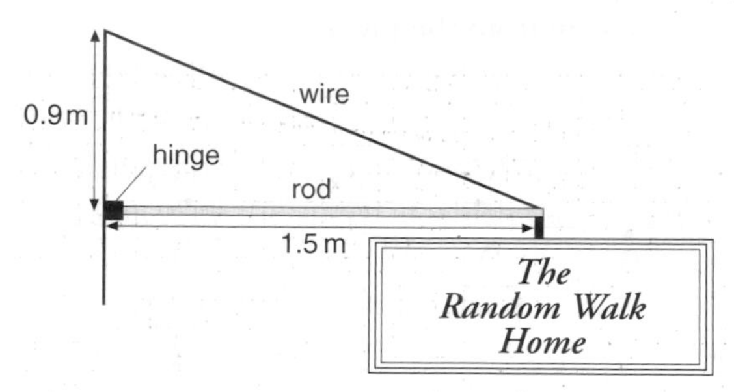

Calculate the clockwise moment produced by the weight of the sign about the hinge. What vertical component of tension is needed at the end of the rod to produce an equal anticlockwise moment? Use the angle that the wire makes with the vertical to find the tension in the wire. Calculate the horizontal component of the tension. Find the horizontal and vertical forces that the hinge must exert to keep the rod and sign in equilibrium.

24.6 A uniform rod AB is 120 cm long and weighs 250 N. It is supported horizontally by two vertical wires, one at A and the other at B. Loads of 100 N and 350 N hang from the rod at distances of 50 cm and 100 cm respectively from A. Calculate the tension in each wire.

Chapter 25

25.1 Define momentum. Is momentum a scalar or a vector quantity?

25.2 Calculate the momentum of (a) a 120 kg rugby player moving at 10 m s^{-1}, (b) an electron (mass 9×10^{-31} kg) moving at 2×10^{7} m s^{-1}, (c) a toy train of mass 1600 g travelling at 25 cm s^{-1}.

25.3 Calculate the change of momentum when (a) a car of mass 800 kg accelerates from 5 m s^{-1} to 30 m s^{-1}, (b) a trolley of mass 0.8 kg slows from 80 cm s^{-1} to 20 cm s^{-1}, (c) a tennis ball of mass 50 g moving horizontally at 7 m s^{-1} hits a vertical wall and rebounds from it at 5 m s^{-1}.

25.4 State Newton's second law in terms of momentum. Show how this version links to the equation $F = ma$.

25.5 If the car in Practice Question 25.3 took 8 s to accelerate, the trolley took 3 s to slow down and the tennis ball was in contact with the wall for 0.6 s, find the forces exerted on them.

Chapter 26

26.1 State the principle of conservation of momentum. Describe how you would show that momentum is conserved in a collision involving two bodies that join together.

26.2 A skateboard of mass 4 kg is moving at a steady speed of 2 m s^{-1} Calculate its momentum. A 1 kg bag of sugar is dropped vertically onto it. Calculate its new speed, assuming the momemtum is unchanged.

26.3 A bullet of mass 20 g moving at 300 m s^{-1} embeds in a stationary wooden block of mass 3980 g. Calculate their combined speed after impact.

26.4 An ice skater, mass 65 kg, moving at 7 m s^{-1}, collides with another skater, mass 45 kg, moving at 6 m s^{-1} in the opposite direction. Calculate their speed and direction as they move off together.

26.5 Explain in terms of momentum conservation how the recoil of a gun arises.

26.6 A boy throws a 100-g stone horizontally at a speed of 6 m s^{-1}. A squirrel, mass 500 g, sitting on ice, catches the stone and then throws it back horizontally at 2 m s^{-1}. Find (a) the initial momentum of the stone, (b) the speed of the squirrel when it has caught the stone, (c) the speed of the squirrel after it has thrown the stone back.

Chapter 27

27.1 Write down a word equation for impulse. What unit of impulse derives directly from this equation? Rewrite this unit in terms of base units only. What other physical quantity has the same unit as impulse?

27.2 A stationary snooker ball of mass 0.15 kg is hit by a cue which exerts an average force of 60 N on it. The duration of the impact is 8 ms. Calculate (a) the impulse of the force exerted by the cue, (b) the speed at which the snooker ball leaves the cue.

27.3 The graph shows how the force applied to a trolley varied with time.

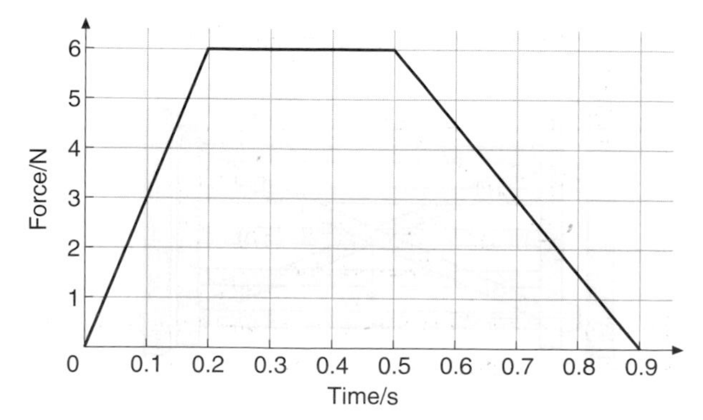

Find the impulse exerted by this force. If the trolley's mass is 800 g, and it started at rest, what was its final speed?

27.4 A car travelling at 30 m s^{-1} crashes through a wall and is brought to rest in 0.5 s. The mass of the car and occupants is 900 kg. Calculate (a) the change in momentum of the car during its collision with the wall, (b) the impulse that the wall exerts on the car during the collision, (c) the average force exerted by the wall on the car during the collision.

27.5 In an experiment to find the average force exerted by a hammer when it is knocking a nail into a block of wood, the following measurements were taken.

Speed of hammer just before it hits the nail 8 m s^{-1}

Speed of hammer rebounding from the nail 2 m s^{-1}

Mass of hammer head 300 g

Time of contact between the hammer and the nail 12 ms

(a) Calculate the change in momentum of the hammer head. (b) What impulse was exerted on the nail by the hammer head? (c) Calculate the average force exerted by the hammer head while it is in contact with the nail. (d) How could you attempt to measure the time of contact in the above experiment?

27.6 Show how Newton's third law can be used to predict that momentum is conserved during a collision.

27.7 A sphere of mass 3 kg is moving at a speed of 4 m s^{-1} towards a stationary sphere of mass 2 kg. After the collision, the 2 kg sphere moves off with a speed of 4.5 m s^{-1}. Calculate the speed of the 3 kg sphere.

Chapter 28

28.1 Define work. Calculate the work done when (a) a force of 60 N moves through a distance of 3 m, (b) a force of 4 kN moves through a distance of 25 cm.

28.2 The graph shows how the force used to stretch a spring varies with extension.

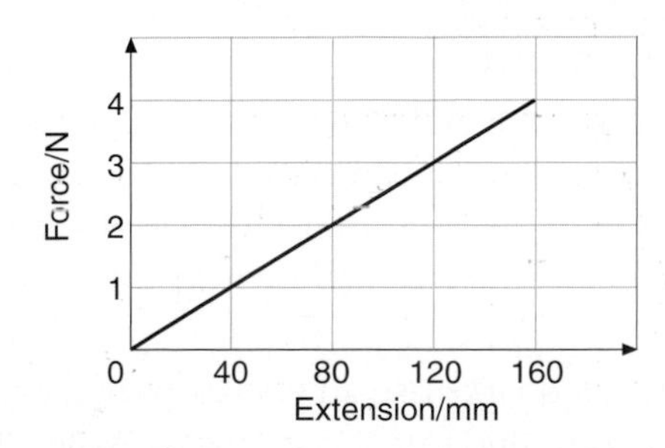

Use the graph to find (a) the work done stretching the spring from its original length to an extension of 60 mm, (b) the work done increasing the extension from 40 mm to 140 mm.

28.3 The graph shows how the force used to stretch a copper wire varies with its extension up to its breaking point.

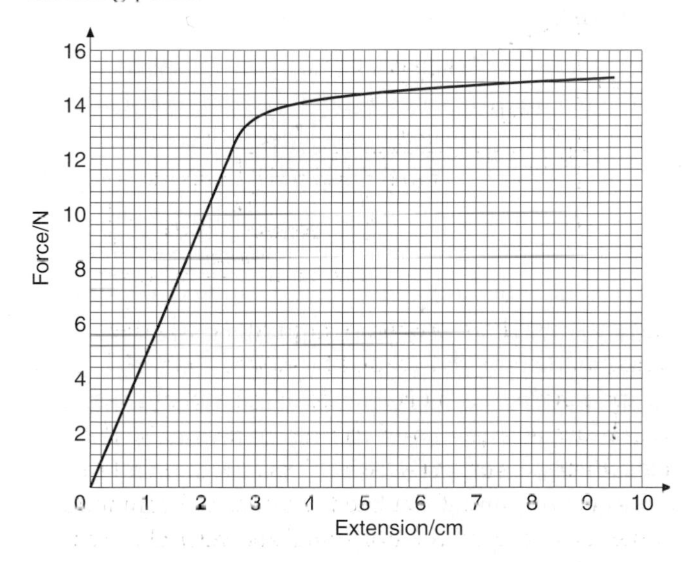

Use the graph to estimate the work done breaking the wire.

28.4 Define power. A man of weight 800 N takes one minute to climb up a staircase 21 m high, travelling a total distance of 36 m in the process. Calculate (a) his average speed, (b) his average upward velocity, (c) the total work he does against his weight, (d) his average power.

28.5 The engine of a car generates a power of 35 kW at a maximum speed of 144 km h^{-1}. Calculate the force acting against the motion of the car at this speed.

Chapter 29

29.1 Is energy a scalar or a vector quantity? Express the unit of energy in terms of base units.

29.2 State the two different forms in which energy can be stored. List three different types of potential energy. Give the more common name for the electromagnetic potential energy associated with a stretched spring.

29.3 Calculate the increase in gravitational potential energy of (a) a football of mass 0.5 kg when it is kicked 25 m into the air, (b) a person of mass 60 kg when they stand up and raise their centre of gravity by 30 cm, (c) a tiddlywink of mass 1.2 mg when flicked 15 cm into the air.

29.4 Calculate the kinetic energy of a car of mass 900 kg moving at 20 m s^{-1}. If this kinetic energy were fully converted into gravitational potential energy, find the vertical height through which the car would rise.

29.5 Repeat Question 29.4 for (a) a trolley of mass 0.6 kg moving at 85 cm s^{-1}, (b) a cricket ball of mass 110 g moving at 40 m s^{-1}. Compare your answer to (b) with the worked example in Chapter 10, page 23.

Chapter 30

30.1 When an object falls at its terminal speed through a fluid, its kinetic energy remains constant while its gravitational potential energy decreases. Explain this apparent loss of energy.

30.2 What is the internal energy of a body? Give three examples of how internal energy may be increased.

30.3 State the principle of conservation of energy.

30.4 A trolley of mass 4 kg is initially at rest on a smooth runway. A string attached to the trolley passes over a frictionless pulley at the bottom of the runway and is tied to a 1 kg mass which is 25 cm above the ground. The trolley is released. Calculate (a) the potential energy lost by the mass falling to the ground, (b) the speed of the trolley and mass just before the mass hits the ground.

30.5 Define efficiency. A 60 W light bulb is 2% efficient as a light source. How much power does it give out as light? What happens to the rest of the energy input?

Chapter 31

31.1 Describe how you would show that momentum is conserved in a collision involving two bodies that do not join together.

31.2 What two quantities are conserved in both elastic and inelastic collisions? What quantity is only conserved in an elastic collision? Give an example of a situation where the collisions are, on average, elastic.

31.3 A railway truck of mass 4×10^4 kg moves at 3 m s^{-1} towards a stationary truck of mass 2×10^4 kg. The trucks collide and join. Calculate the total momentum of the two trucks before the collision. What is the total momentum of the two trucks after the collision? Find the speed at which the joined trucks move off after their collision. Compare the kinetic energy after the collision with that before the collision. Why does your answer not contradict the law of conservation of energy?

31.4 A pendulum is pulled to one side and released. State the energy changes that occur during one complete swing.

31.5 A bullet of mass 20 g is moving at a speed of 300 m s^{-1} towards a stationary wooden block of mass 3980 g. The bullet embeds in the block. (a) Calculate the momentum of the bullet before the collision. Why is this equal to the momentum of the bullet and block after the collision? Calculate the combined speed after they collide. (b) Use your answers to part (a) to calculate the kinetic energy of the bullet before the collision, and the kinetic energy of the combined bullet and the block after the collision. Comment on your answers.

31.6 Explain why motorway crash barriers are designed to give highly inelastic collisions with any vehicle that hits them.

Chapter 32

32.1 List, in order of decreasing mass, the three principal particles that make up an atom. For each particle give the sign of its charge.

32.2 Describe the structure of an atom. Sketch this structure for an atom of beryllium-8, $^{8}_{4}$Be.

32.3 Explain why the density of nuclear matter is very much greater than the overall density of the material that it forms a part of.

32.4 The symbol $^{207}_{82}$Pb represents an atom of lead-207. State (a) the number of protons in its nucleus, (b) the number of nucleons in its nucleus, (c) the number of neutrons in its nucleus.

32.5 What is an isotope of an element? Tin has 25 isotopes. The lightest is $^{108}_{50}$Sn. Write down the symbols for three possible heavier ones.

Chapter 33

33.1 Draw a diagram of the experimental arrangement used in the alpha particle scattering experiment. Describe the observations that were made.

33.2 With the aid of a sketch, explain how a material with a nuclear atomic structure would lead to the observations made in the alpha particle scattering experiment.

33.3 Approximately how many times bigger is the diameter of an atom compared to that of a nucleus? A head-on view of a spherical atom is a circle. Approximately what percentage of the area of this circle is taken up by the nucleus?

33.4 What is a quark? Use an encyclopaedia or an Internet search to find out about quarks. Write a few paragraphs to describe what you have found out.

33.5 Compare deep inelastic scattering in terms of the particles used (both incident and target), the process involved and the results achieved with alpha particle scattering.

Chapter 34

34.1 What happens to the atoms when air is ionised? State two methods by which air can be ionised.

34.2 Describe how you would use an ionisation chamber to measure how far the alpha particles emitted from a radioactive source can travel through air.

34.3 Polonium-215 ($^{215}_{84}Po$) and actinium-227 ($^{227}_{89}Ac$) both emit alpha particles. Write equations for their decays.

34.4 A smoke detector consists of an alpha source and an alpha particle detector separated by an air gap. Draw a diagram of this arrangement. Explain what change would take place if smoke came into the space between the source and the detector.

34.5 Thorium-228 ($^{228}_{90}Th$) decays by alpha emission to form an isotope of radium (symbol Ra). Write down the complete decay equation for this process. After how many more alpha decays will the isotope lead-212 be produced? What is the proton number of lead-212?

Chapter 35

35.1 What advantages does a Geiger–Müller tube have over an ionisation chamber? Are there any disadvantages?

35.2 Describe how you would use a GM tube, a counter and a method involving penetrating power to check that a source emits only beta particles.

35.3 A student is confused when told that beta-minus decay involves the ejection of an electron from the nucleus of an atom, as he knows that all nuclei contain only protons and neutrons and do not contain any electrons. Explain to him what has actually happened.

35.4 $^{216}_{84}Po$ (polonium-216) decays to lead (Pb) with the emission of an alpha particle. The lead then emits gamma radiation before decaying to bismuth (Bi) with the emission of a beta-minus particle. Write down the nuclear equation for each of these three decays.

35.5 $^{241}_{94}Pu$ (plutonium-241) decays via β^-, α, α, β^-, α, α, β^-, α, α, α, β^-, α, β^- decays to $^{209}_{83}Bi$ (bismuth). Illustrate this decay chain by drawing a figure similar to Figure 35.4, page 75.

Chapter 36

36.1 Radioisotopes contribute to *background radiation*. These isotopes can have *natural radioactivity* or *artificial radioactivity*. Explain the three terms in italics. Give another important contributor to background radiation other than radioisotopes.

36.2 In what way is radioactive decay a random process? How, despite this, is it still possible to use mathematics to predict the pattern that the decay will follow?

36.3 Define the terms activity and decay constant. Calculate the activity of a radioactive isotope whose decay constant is $8.0 \times 10^{-6}\ s^{-1}$ at the instant when there are 3.0×10^{11} atoms of that isotope present. Explain why this activity will decrease with time.

36.4 Define half-life. Describe the experiment that you would use to determine the decay constant of a radioactive source that is known to have a half-life of about a minute.

36.5 The decay constant of a radioactive isotope of strontium is $7.84 \times 10^{-10}\ s^{-1}$. Show that its half-life is approximately 28 years. A source initially contains 4.5 mg of this strontium isotope. Calculate the mass of this isotope that will remain after 84 years.

Assessment questions

The following questions have been chosen or modified to be similar in style and format to those which will be set for the AS assessment tests. These questions meet the requirements of the assessment objectives of the specification.

1 Classify each of the terms in the left-hand column by placing a tick in the relevant box.

	Base unit	Derived unit	Base quantity	Derived quantity
Length				
Kilogram				
Power				
Joule				

(Total 4 marks)

(Edexcel GCE Physics Module Test PH1, June 1998)

2 Complete each of the following statements with a single word:

The rate of change of displacement is called …………..

The rate of doing work is called …………

The rate of change of momentum is called …………

(Total 3 marks)

(Edexcel GCE Physics Module Test PH1, June 1998)

3 Two cars, A and B, are travelling along the outside lane of a motorway at a speed of 30.0 m s^{-1}. They are a distance *d* apart.

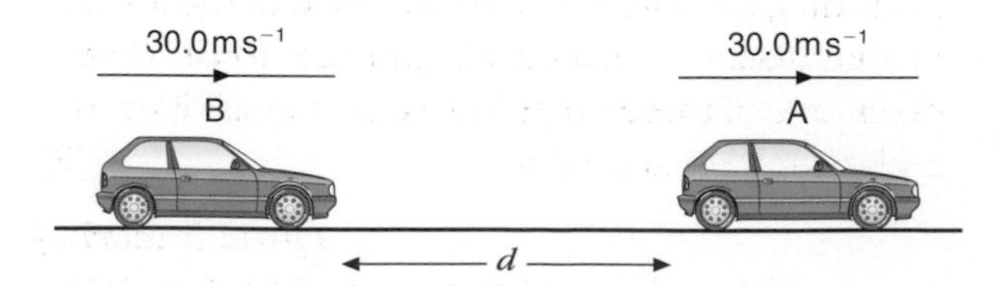

The driver of car A sees a slower vehicle move out in front of him, and brakes hard until his speed has fallen to 22.0 m s^{-1}. The driver of car B sees car A brake and, after a reaction time of 0.900 s, brakes with the same constant deceleration as A. The diagram shows velocity–time graphs for car A (solid line) and car B (broken line).

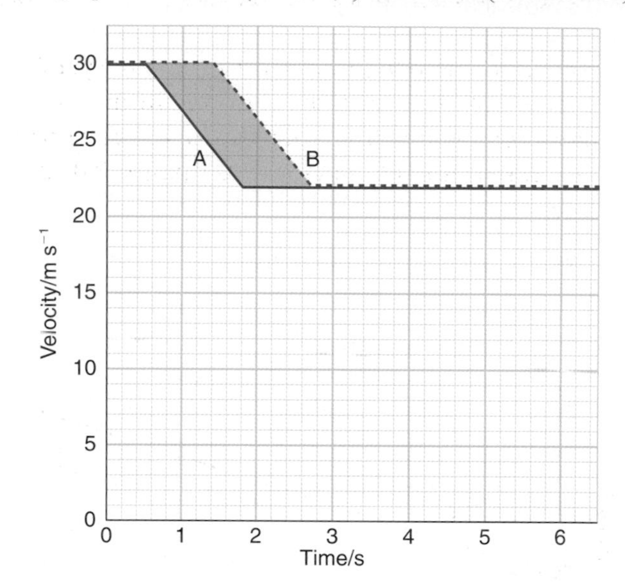

Find the deceleration of the cars whilst they are braking. **[3]**

What does the area under a velocity-time graph represent? **[1]**

Determine the shaded area. **[2]**

State the minimum value of the initial separation *d* if the cars are not to collide. Explain how you arrived at your answer. **[2]**

Suppose that, instead of only slowing down to 22.0 m s^{-1}, the cars had to stop. Add lines to the grid to show the velocity-time graphs in this case. (Assume that the cars come to rest with the same constant deceleration as before.) **[1]**

Explain why a collision is now more likely. **[2]**

(Total 11 marks)

(Edexcel GCE Physics Unit PHY1, June 2002)

4 The diagram below shows a trolley running down a slope.

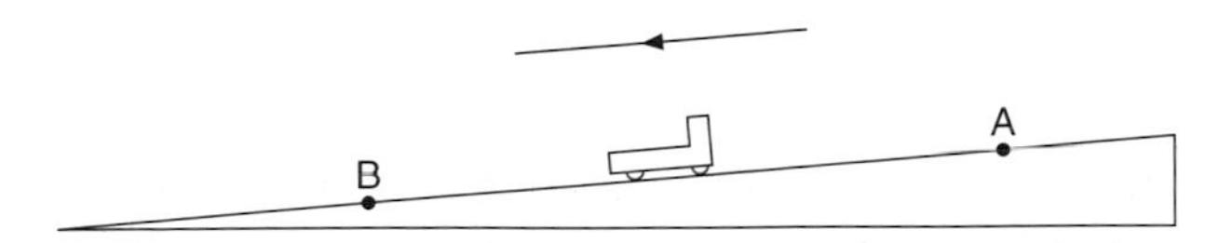

Complete the diagram to show an experimental arrangement you could use to determine how the trolley's position varies with time. **[2]**

The data is used to produce a velocity–time graph for the trolley. Below is the graph for the motion from point A to point B. Time is taken to be zero as the trolley passes A, and the trolley passes B 0.70 s later.

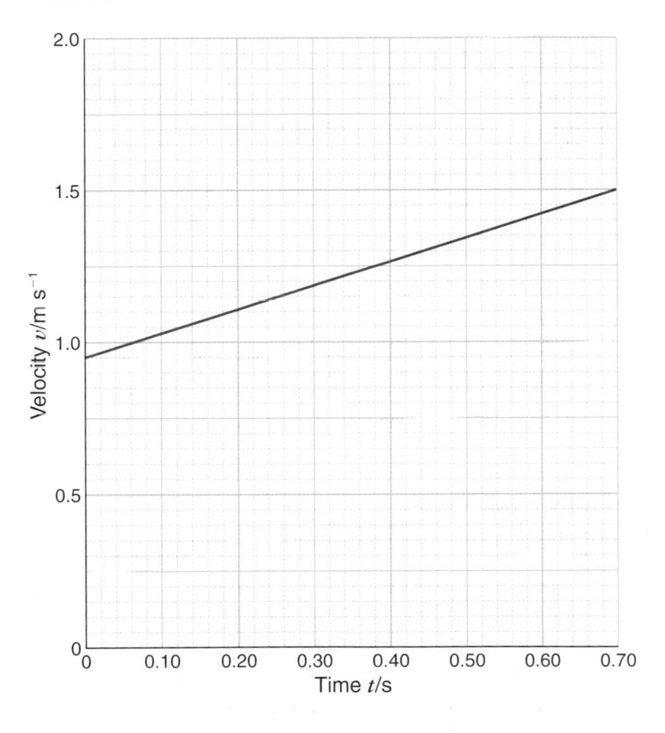

The motion shown on the graph can be described by the equation $v = u + at$. Use information from the graph to determine values for u and a. **[3]**

Determine the distance AB. **[3]**

Sketch a graph to show how the displacement x of the trolley from point A varies with time t. Add a scale to each axis. **[3]**

(Total 11 marks)

(Edexcel GCE Physics Unit PHY1, January 2002)

5 Write down a word equation which defines the magnitude of a force. **[2]**

Two forces have equal magnitudes. State three ways in which these two equal forces can differ. **[3]**

(Total 5 marks)

(Edexcel GCE Physics Module Test PH1, January 1998)

6 State Newton's second law of motion. **[2]**

You are asked to test the relation between force and acceleration. Draw and label a diagram of the apparatus you would use. State clearly how you would use the apparatus and what measurements you would make. **[2, 4]**

Explain how you would use your measurements to test the relationship between force and acceleration. **[3]**

(Total 11 marks)

(Edexcel GCE Physics Module Test PH1, June 1997)

7 When the jet engines on an aircraft are started, fuel is burned and the exhaust gases emerge from the back of the engines at high speed. With reference to Newton's second and third laws of motion, explain why the aircraft accelerates forward. You may be awarded a mark for the clarity of your answer.

(Total 4 marks)

(Edexcel GCE Physics Unit PHY1, January 2002)

8 A magnet X is clamped to a frictionless table. A second magnet Y is brought close to X and then released.

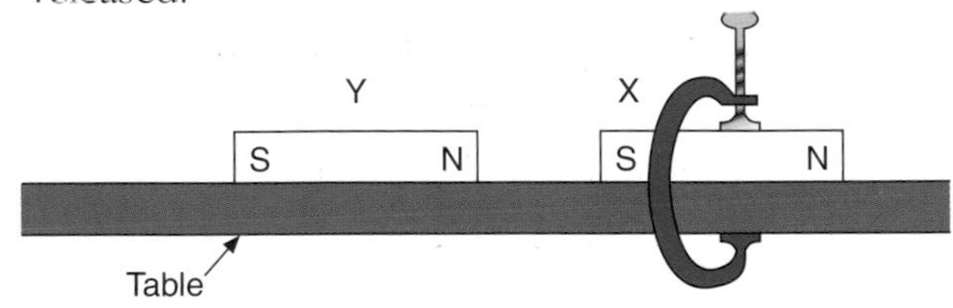

Add labelled forces to the free-body diagram below for magnet Y to show the forces acting on it just after it is released.

[3]

According to Newton's third law, each of the forces in your diagram is paired with another force. Write down one of these other forces, stating its direction and the body it acts upon. **[2]**

(Total 5 marks)

(Edexcel GCE Physics Unit PHY1, June 2002)

ASSESSMENT QUESTIONS

16 State the composition of an alpha particle. **[1]**

When an alpha particle passes through matter, it may ionise atoms. Explain what **ionise** means. **[1]**

An alpha particle from a certain radioactive source has a kinetic energy of 8.2×10^{-13} J. Using the information below, estimate how long it would take this alpha particle to travel a distance equal to the diameter of an atom.
Mass of alpha particle = 6.6×10^{-27} kg
Diameter of an atom = 1.0×10^{-10} m **[3]**

A beta particle from a different radioactive source has the same kinetic energy as the alpha particle. Explain qualitatively how the speed of this beta particle would compare with the speed of the alpha particle. **[2]**

Beta particles are many times less effective at ionising atoms than alpha particles. Suggest a reason for this. **[1]**

(Total 8 marks)
(Edexcel GCE Physics Unit PHY1, June 2002)

17 Geiger and Marsden carried out a scattering experiment, which led to a revised understanding of the structure of the atom. The following tables refer to this experiment. Copy and complete the tables and the sentences that follow them.

	Name
Incoming particle	
Target atoms	

[2]

Observation	Conclusion about atomic structure
The incoming particles were mostly deflected	
A few particles were deflected by angles greater than 90°.	

[2]

The diameter of .... is approximately 10^{-15} m.

The diameter of .... is approximately 10^{-10} m. **[2]**

(Total 6 marks)
(Edexcel GCE Physics Module Test PH2, June 1999)

18 Compare and contrast deep inelastic scattering with alpha particle scattering **[2, 2, 2]**

(Total 6 marks)
(Edexcel GCE Physics Module Test PH4, January 1998)

19 Samples of two different isotopes of iron have been prepared. Compare the composition of their nuclei. **[2]**

The samples have the same chemical properties. Suggest a physical property which would differ between them. **[1]**

Tritium (hydrogen-3) is an emitter of beta-minus particles. Complete the nuclear equation below for this decay. **[3]**

H → He + β

Describe how you would verify experimentally that tritium emits only beta particles. **[4]**

(Total 10 marks)
(Edexcel GCE Physics Unit PHY1, January 2002)

20 A school physics department has a cobalt-60 gamma ray source. Its half-life is several years. As a project, a student decides to try to measure this half-life. She sets up a GM tube close to the source and determines the count produced by the source in a 10 minute period. One year later she repeats the measurement. Explain **two** precautions, other than safety precautions, which the student should take in her measurements in order to produce a reliable value for the half-life. **[3]**

The student then calculates the count she would expect to get in a 10 minute period if she repeated her measurement annually. Some of her results are shown in the table below.

Time/year	Count in 10 minute period
0	17 602
1	15 489
2	13 630
3	
4	10 554
5	9287
6	8172
7	7191

Calculate the ratio of the count after one year to the initial count. **[1]**

Complete the table to show the count you would expect for year 3. **[1]**

Plot a graph of count against time and use your graph to determine a value for the half life of cobalt 60. **[4]**

(Total 9 marks)

(Edexcel GCE Physics Unit PHY1, June 2002)

21 A radioactive source contains barium-140. The initial activity of the source is 6.4×10^8 Bq. Its decay constant is 0.053 day^{-1}. Calculate the half-life, in days, of barium-140. **[2]**

Calculate the initial number of barium-140 nuclei present in the source. **[2]**

The graph below represents radioactive decay. Add a suitable scale to each axis, so that the graph correctly represents the decay of this barium source.

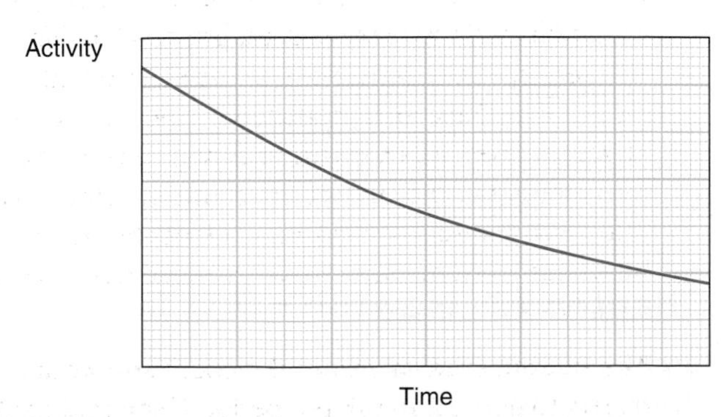

A radium-226 source has the same initial activity as the barium-140 source. Its half-life is 1600 years. On the same axes, sketch a graph to show how the activity of the radium source would vary over the same period. **[3]**

(Total 7 marks)

(Edexcel GCE Physics Unit PHY1, January 2002)

22 Complete the following equations by writing the correct physical quantities, in words, on the dotted lines.

$$\text{volume} = \frac{\text{mass}}{\dots\dots\dots}$$

$$\text{average velocity} = \frac{\dots\dots\dots}{\text{time}}$$

$$\frac{\text{weight}}{\text{mass}} = \dots\dots\dots$$

$$\text{decay constant} = \frac{\text{activity}}{\dots\dots\dots}$$

(Total 4 marks)

(Edexcel GCE Physics Unit PHY1, January 2002)

23 Five graphs, A–E, are shown below.

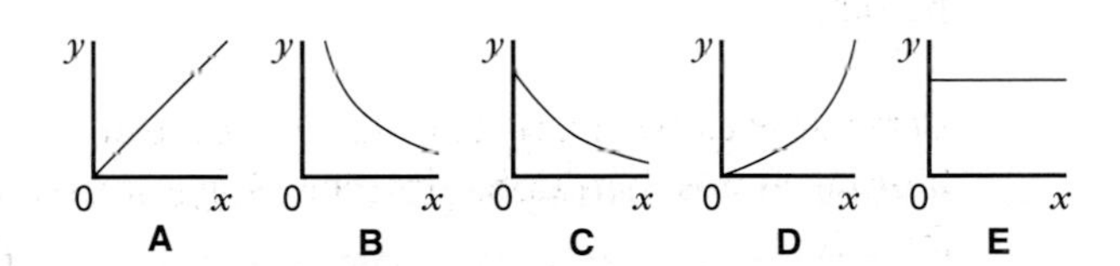

Write the appropriate letter in the table below to indicate which graph is obtained when the quantities described below are plotted on the *y* and *x* axes.

Variable on y axis	Variable on x axis	Graph
Activity of a radioactive source	Time	
Increase in gravitational potential energy of a body	Vertical height body is raised	
Acceleration produced by a constant force	Mass of body being accelerated	
Half-life of a radioactive source	Number of nuclei present in the source	

(Total 4 marks)

(Edexcel GCE Physics Unit PHY1, June 2002)

Things you need to know

Chapter 1 Making measurements precisely

resolution: smallest difference in a reading that an instrument can indicate

vernier callipers: instrument used to measure distances to a resolution of 0.1 mm

micrometer screw gauge: instrument used to measure distances to a resolution of 0.01 mm

uncertainty: range around the measured reading within which the true reading lies

precision: a precise measurement has a low uncertainty

Chapter 2 Solids, liquids and gases: some measurable properties

density: mass per unit volume

rigid: something that keeps its shape

fluid: something that flows (a liquid or a gas)

Chapter 3 The International System of Units

physical quantity: physical property that can be measured

base quantities: agreed minimum set of starting quantities for the system of measurement

unit: agreed quantity, used for comparing other quantities

base unit: unit for a base quantity

Chapter 4 Derived quantities and units

derived quantity: quantity derived from combinations of base quantities

derived unit: unit for a derived quantity

homogeneous: of the same type; correct equations must be homogeneous, but homogeneous equations need not be correct

Chapter 5 Displacement and velocity

displacement: distance moved in a particular direction

scalar: quantity having size only and no direction

vector: quantity having size and direction

speed: distance/time; a scalar quantity

average speed: total distance/total time

velocity: displacement/time; a vector quantity

instantaneous velocity: velocity at that instant; found by dividing the change in displacement by the very short time interval over which it is measured

Chapter 6 Acceleration

acceleration: change in velocity/time; a vector quantity

rate of: divided by time

Chapter 7 Motion graphs

gradient of displacement–time graph: (instantaneous) velocity

gradient of velocity–time graph: (instantaneous) acceleration

area under velocity–time graph: change in displacement

Chapter 11 Horizontal projection

projectile motion: horizontal and vertical motions are independent of each other; horizontal velocity constant if air resistance is ignored; vertically the body accelerates downwards all the time

Chapter 12 Newton's first law

force: something that can cause a body to accelerate; a vector quantity

equilibrium: condition where the resultant force on a body is zero; any forces acting are balanced

resultant force: single force that could replace all other forces and have the same effect

Newton I: a body will remain at rest or continue to move with a constant velocity as long as the forces on it are balanced

inertia: reluctance of a body to change its velocity

Chapter 13 Newton's second law

one newton: the resultant force that will give a mass of 1 kg an acceleration of 1 m s^{-2}

Newton II: acceleration of a body is proportional to the resultant force and takes place in the direction of the force

Chapter 15 Types of forces

weight: gravitational force on a body

contact forces: electrostatic forces between the outer layers of electrons of the bodies in contact

Chapter 16 Free-body force diagrams – 1

Newton III: while a body A exerts a force on body B, body B exerts a force on body A. The forces are equal, opposite and of the same type; they have the same line of action and act for the same time

Chapter 19 Components of a force

component: effect of a vector in a particular direction; magnitude = force × cosine of angle between original and required direction

Chapter 20 Fluid forces

upthrust: upward force acting on an immersed object due to the difference in the fluid pressure acting on its top and its bottom

terminal speed (terminal velocity): the steady speed attained by an object falling through a fluid

drag: opposition of a fluid to the movement of an object through it

aerodynamic lift: force lifting a wing caused by the pressure of moving air on top of wing being less than that underneath

Chapter 21 Free-body force diagrams – 3

tangential force: force that acts along a surface

Chapter 22 Moments and couples

tension: state caused by forces that stretch an object

compression: state caused by forces that squash an object

couple: pair of equal and opposite parallel forces that are not in line and produce a rotational effect

moment of a couple: one of the forces × perpendicular distance between them

moment of a force: force × perpendicular distance from a point

torque: turning moment of a set of forces that tend to cause rotation

principle of moments: if a body is in equilibrium, the sum of the moments about any point must be zero

Chapter 23 Conditions for equilibrium

centre of gravity (mass): point where all the weight of the body appears to act

Chapter 25 Momentum

momentum: mass × velocity

Newton II: rate of change of momentum of a body is directly proportional to the resultant force acting on it and takes place in the same direction as the resultant force

force: quantity which causes a rate of change of momentum

Chapter 26 Conservation of momentum

conservation of momentum: provided no external forces act, the total momentum in any direction remains constant

Chapter 27 Impulse

impulse: force × time; area under force–time graph; equal to change of momentum

Chapter 28 Work, energy and power

work: force × distance moved in the direction of the force; area under force–displacement graph; equal to the energy transferred from one system to another

system: any body or group of bodies under consideration

energy: the property of a sytem that indicates its ability to do work; when body A works on body B, the amount of energy transferred from A to B is equal to the work done.

power: rate of doing work: force × velocity; watt = joule/second

Chapter 29 Potential and kinetic energy

gravitational potential energy: energy associated with position in a gravitational field; change = $mg\Delta h$

kinetic energy: energy associated with the motion of a body = $\frac{1}{2}mv^2$

Chapter 30 Conservation of energy

internal energy: random kinetic and potential energy of the molecules of a body

conservation of energy: energy content of a closed or isolated system remains constant

efficiency: useful energy output/total energy input

Chapter 31 Elastic and inelastic collisions

elastic collision: one in which kinetic energy is conserved

inelastic collision: one in which kinetic energy is not conserved

Chapter 32 Inside the atom

nucleus: tiny positive centre of an atom in which most of its mass is concentrated

proton number (or **atomic number**) Z**:** number of protons in the nucleus and the number of electrons in a neutral atom

nucleon: collective name for particles found in the nucleus, namely protons and neutrons

nucleon number (or **mass number**) A**:** total number of protons Z and neutrons N in the nucleus; indicates mass of atom

isotopes: atoms with the same number of protons but different numbers of neutrons in their nuclei

Chapter 33 Scattering

alpha particle (Rutherford) scattering: elastic collisions between the alpha particles and the nucleus, that reveal the size of the nucleus

quarks: particles found in both protons and neutrons

deep inelastic scattering: high energy particles being deflected and losing energy after inelastic collisions with the internal structure of another elementary particle

Chapter 34 Alpha radiation

ionises: atom given sufficient energy to release an electron and become a positive ion

alpha (α) **radiation:** helium nuclei emitted by some radioactive nuclei; heavily ionising (produces many ions per millimetre along its path) and short range

Chapter 35 Beta and gamma radiations

beta-minus (β^-) **radiation:** fast-moving electrons emitted by some radioactive nuclei; less ionising and consequently having a larger range than alpha

beta-plus (β^+) **radiation:** fast-moving positrons emitted by some radioactive nuclei

gamma (γ) **rays:** photons of electromagnetic radiation emitted from energetic nuclei resulting from alpha or beta decay; little ionising and large range

Chapter 36 Rate of decay

background radiation: emissions mainly from naturally occurring radio-isotopes; varies randomly with time

random decay: unable to predict when any given nucleus will decay, although large numbers of nuclei can be treated statistically

activity: the number of decays of a radioactive source per second

decay constant: probability of decay per nucleus per second

half-life: average time taken for half the nuclei of that isotope to decay

Equations to learn

Weight	weight = mg
Velocity	$v = \frac{\Delta x}{\Delta t}$
Acceleration	$a = \frac{\Delta v}{\Delta t}$
Momentum	$p = mv$
Force	$F = ma$
Work done	$\Delta W = F\Delta x$
Power	$P = \frac{\Delta W}{\Delta t}$
Kinetic energy	$KE = \frac{1}{2}mv^2$
Gravitational potential energy	$\Delta W = mg\Delta h$ (close to the Earth)

Index

INDEX